Urban Public Transport Systems Innovation in the Fourth Industrial Revolution Era

Trynos Gumbo · Thembani Moyo ·
Bongumusa Ndwandwe · Brightnes Risimati ·
Siphiwe Given Mbatha

Urban Public Transport Systems Innovation in the Fourth Industrial Revolution Era

Global South Perspectives, Reflections and Conjectures

Springer

Trynos Gumbo
Sustainable and Smart Cities and Regions Research Group, Department of Urban and Regional Planning
University of Johannesburg
Johannesburg, Gauteng, South Africa

Bongumusa Ndwandwe
Sustainable and Smart Cities and Regions Research Group, Department of Urban and Regional Planning
University of Johannesburg
Johannesburg, Gauteng, South Africa

Siphiwe Given Mbatha
Sustainable and Smart Cities and Regions Research Group, Department of Urban and Regional Planning
University of Johannesburg
Johannesburg, Gauteng, South Africa

Thembani Moyo
Sustainable and Smart Cities and Regions Research Group, Department of Urban and Regional Planning
University of Johannesburg
Johannesburg, Gauteng, South Africa

Brightnes Risimati
Department of Urban and Regional Planning
University of Venda
Thohoyandou, Limpopo, South Africa

Sustainable and Smart Cities and Regions Research Group, Department of Urban and Regional Planning
University of Johannesburg
Johannesburg, Gauteng, South Africa

Funded by National Research Foundation (NRF) South Africa, Thuthuka Funding Instrument (NRF Rating Track) Granted to Prof Trynos Gumbo; Reference: TTK160612170527

ISBN 978-3-030-98716-9 ISBN 978-3-030-98717-6 (eBook)
https://doi.org/10.1007/978-3-030-98717-6

This Springer imprint is published by the registered company Springer Nature Switzerland AG
The registered company address is: Gewerbestrasse 11, 6330 Cham, Switzerland

This book is dedicated to all urban public transport commuters, particularly in most cities of the developing world who toil every day on their journeys to and from work, shopping, and many other trips and yearn for physically and electronically integrated, secure, safe, affordable, and convenient transport systems.

Foreword

When considering the dynamic and fast-paced nature of modern cities, the need for ensuring integrated and well-coordinated development and movement within the cities has never been so imperative. The ever-increasing challenges associated with urbanization and development pose an immerse demand for well thought out and researched planning and development proposals to ensure the success of cities in the future. Modern cities in Southern Africa have made extensive investments in improving urban mobility. This comprises all-inclusive transportation infrastructure instead of only focusing on one part of the journey through facilitating spatial integration between diverse transport modalities. Innovations can see this in implementing the transport infrastructure and services such as the newly built Bus Rapid Transit, Rail Rapid Transit, cycling lanes, and improved pedestrians' walkways. Given these developments, this book ponders the question of whether these innovations and investments in transport infrastructure and services contribute to the creation of a holistic public transport system.

Urban mobility literature has outlined, the problem of uncoordinated development is unprecedented in rapidly growing cities in the global south. This is predominantly within the capital cities of most African countries that have been experiencing exponential growth of both populations and their spatial coverage as they continuously extend outwards. Acknowledging this reality and the role that good planning can play in ensuring functional and well-performing public transportation systems, this book embarked on an exercise to unpack the trends of public transportation development within cities in Southern Africa. This is undertaken through numerous studies that assess policy frameworks, legislature, and case studies of selected cities.

One key feature of the book is to acknowledge urban public transportation in developing cities. However, there remain yawning gaps in integrating the systems to promote convenience for commuters. As the innovations in urban public transit are still at their initial implementation and use phases, the book contributes to the existing literature by unpacking the current state of integration among the various public transport systems. It reveals how working relationships could be improved, as the projects are rolled out further to reach all community sections. To this extent, the book calls for holistic urban mobility planning and designs frameworks that promote integrated multiplicity in travel opportunities and achieve smart mobility. I am persuaded that this book is a valuable resource for activists, academics, policymakers, politicians, and urban planners in the needed engagement on the subject under consideration.

Prof. James Chakwizira
Department of Urban and Regional Planning
North West University
Potchefstroom, South Africa

Preface

This book explores the physical and electronic integration of innovative urban public transport systems in seven metropolitan cities in South Africa and Zimbabwe in the era of the Fourth Industrial Revolution (Industry 4.0). The book also highlights how collaborative engagement can improve new transport projects in cities of the Global South. It demonstrates how integration concerns remain in transport infrastructure projects in cities of the developing countries. Consequently, in order to strengthen the emerging and promising economies of these cities, there is a need for efficient, integrated, reliable and affordable public transport systems. The book explains that plans to deliver innovative transport systems in the Global South need to be well coordinated and managed to yield physically and electronically integrated systems.

Johannesburg, South Africa

Trynos Gumbo
Thembani Moyo
Bongumusa Ndwandwe
Brightnes Risimati
Siphiwe Given Mbatha

Acknowledgments

The idea of this book was conceived in 2019 after the completion of the National Research Foundation, Thuthuka funded project investigated innovative transport systems in Gauteng province. The fund had also supported masters' students who had researched the topic, hence the co-authorship with my former students. We are truly grateful to everyone who contributed to this book, including the National Research Foundation (NRF), the University of Johannesburg (UJ), the Johannesburg Metropolitan Municipality (JMM), the Tshwane Metropolitan Municipality (TMM), the Ekurhuleni Metropolitan Municipality (EMM) City of Cape Town (CCT); City of Harare (COH); City of Bulawayo (COB); public transport operators, municipal officials, transport experts, commuters and relevant national and provincial officials. We also acknowledge the inputs of the anonymous peer reviewers who helped to shape the ideas in this book as well as the editor, who guided this project from its proposal stage until its fruition. It is also important to recognize all the authors who made efforts to respond to all questions and include all the suggestions in this book. Relevant data, materials, critical perspectives, and insights were crystallized to give this volume. The authors take responsibility for any omissions, oversights, and mistakes, hence any requests for corrections have been welcome at any given time.

Contents

About the Authors

Prof. Trynos Gumbo is a professional planner and currently an Associate Professor and Head of the Department of Urban and Regional Planning within the Faculty of Engineering and the Built Environment at the University of Johannesburg (UJ). He Holds a Ph.D. from Stellenbosch University, South Africa as well as master's and honors degrees from the University of Zimbabwe (UZ), Zimbabwe. He has previously worked in the Africa Institute of South Africa of the Human Sciences Research Council as a research specialist and Acting Head for the sustainable development program. Prof Gumbo has also worked as an international instructor in the urban management master's program within the Ethiopian Civil Service University College (ECSUC) in Addis Ababa in Ethiopia. Before, Prof Gumbo had worked as lecturer and Head of Department at the National University of Science and Technology (NUST) in Zimbabwe. He has attended and presented at several national and international conferences and has published widely in a variety of research areas that include informality, housing, urban planning, development, and management. His research interests include urban transportation planning and management, sustainable and smart cities development, housing and economic informality, green economy, and renewable energy generation from waste and innovative building technologies and materials.

Mr. Thembani Moyo recently completed his doctoral studies and is currently a Post Doctoral Fellow at the Centre for Applied Research and Innovation in the Built Environment (CARINBE) and also a part time lecturer in the Department of Urban and Regional Planning, Faculty of Engineering and the Built Environment, University of Johannesburg, Republic of South Africa. He is also a part-time lecturer at the Department of Urban and Regional Planning. Prior to this, he graduated with a B.Tech. (Hons.) in Town and Regional in 2014 and in 2016 he obtained an M.Tech. degree in Operations Management at the University of Johannesburg, Republic of South Africa. His research interests leverage a combination of experimental and empirical approaches to study complex problems in the field of urban planning at the interface of Geographical Information Systems and Remote Sensing. In addition, providing quality solutions to complex information system problems with scope, level, and tight schedules, in a bid to continue making a significant contribution within the pursuit of the merging the Fourth Industrial Revolution (4IR) and urban planning. He has published in reputable chapters, peer-reviewed journals, and peer-reviewed conference proceedings.

Mr. Bongumusa Ndwandwe is a Professional Town Planner. He graduated with a B.Tech. (Hons.) in Town and Regional in 2013 and in 2018 he obtained an M.Tech. degree in Operations Management at the University of Johannesburg, Republic of South Africa. He is inspired by the desire to promote development in the community and to empower those less fortunate in our midst. This compelling desire to effect change and to contribute to community development has prompted him to study how cities and regions function in the Global South. In future, he intends on expanding his knowledge of development through studying Project Management, Development Finance, and Law. Such knowledge would equip him with the needed skills to be more effective in his approach to community development. He has published several research papers in the areas of transportation planning and smart cities. Through his various engagements in community activities and projects, his willingness to learn and leadership qualities have been apparent.

Mr. Brightnes Risimati is a Professional Planner and NGAP Lecturer in the Department of Urban and Regional Planning within the Faculty of Science, Engineering and Agriculture, University of Venda (Univen) and is associated with the Sustainable and Smart Cities and Regions Research Group. He is a former Lecturer in the Department of Urban and Regional Planning within the Faculty of Engineering and the Built Environment at the University of Johannesburg (UJ). Mr. Risimati holds a Master of Urban and Regional Planning, Master of Technology Degree in Operations Management, Bachelor of Technology degree, and Diploma in Town and Regional Planning all from University of Johannesburg, South Africa. In partnership with XIVILA Development Group (Pty) Ltd and Emendo (Pty) Ltd, Mr. Risimati is actively involved in a variety of Spatial Planning and Land-Use Management projects in South Africa including assessment, categorization, and development of informal settlements upgrading plans in the City of Johannesburg Metropolitan Municipality, Gauteng Province. His research interests include urban transportation planning and management, sustainable and smart cities development, informal settlement upgrading, Land Economics, and Tenure Systems.

Mr. Siphiwe Given Mbatha is an Assistant Lecturer in the Department of Urban and Regional Planning, Faculty of Engineering and Built Environment, University of Johannesburg, Republic of South Africa. He graduated with B.Tech. (Hons.) in Town and Regional in 2017 and in 2019 completed an M.Tech. degree in Operations Management at the University of Johannesburg, Republic of South Africa. His research interest includes urban public transport planning and management, housing, and informal settlement. He has attended and presented at national and international conferences and has published numerous conference papers on urban public transport planning and management.

Acronyms and Abbreviations

4IR	Fourth Industrial Revolution
BRT	Bus Rapid Transit
CBD	Central Business District
EMM	Ekurhuleni Metropolitan Municipality
ITS	Intelligent Transportation Systems
JMM	Johannesburg Metropolitan Municipality
NMT	Non-motorized transportation
NRF	National Research Foundation
O-D	Origin and Destination
PT	Public transport
PTS	Public Transport System
RRT	Railway rapid transit
TDM	Travel demand management
TMM	Tshwane Metropolitan Municipality
UJ	University of Johannesburg
UPT	Urban Public transport
COB	City of Bulawayo
CCT	City of Cape Town
COH	City of Harare

Part I
Setting the Scene

Chapter 1
Urban Public Transport Systems Innovations in the Era of the Fourth Industrial Revolution: Perspectives, Reflections, and Conjectures

Keywords Innovative urban transport · Smart technology · Planning · Integration · Southern African cities

1.1 Introduction

The planning and provision of efficient transport systems remain a pipe dream for many in cities of the developing world. Most urban centers in developing countries are faced with shortages of reliable transport systems particularly in the Asian, Latin American, and African continents (Mishra, 2019; Ojo, 2019). It has been noted that the problem is unprecedented in rapidly growing cities, and predominantly within the capital cities of most African countries. These have experienced exponential growth of both population and spatial coverage as they continuously extend outwards. Such growth trends have also occurred in Southern African large urban centers. Besides the positive outcomes of rapid urbanization currently experienced within cities in Southern Africa, there have been negative intensive pressures exerted on the cities, forcing innovative responses to these challenges (Gumbo & Moyo, 2020). This has led to a growth in formal and informal innovative public transportation systems in response to mobility demands. Formal innovative public transportation systems include bus rapid transportation (Scorcia & Munoz-Raskin, 2019); rail rapid transportation (Arnold et al., 2017; Fombad, 2015; Moyo & Musakwa, 2016a; Van Der Westhuizen, 2007); rickshaws (Bokopane et al., 2015; Mbara, 2016) and carpooling services (Cramer & Krueger, 2016; Henama & Sifolo, 2017). Informal systems include the mini-bus taxis (Govender & Allopi, 2006; Lomme, 2008), and pirate taxis (Dumba, 2017; Mbara, 2016; Mbara & Pisa, 2018).

Characteristic of rapidly growing urban centers; the cities in Southern Africa are currently facing high transport demands which have surpassed the capacity of individual municipalities. The pressure to devise sustainable, efficient public transport

T. Gumbo et al., *Urban Public Transport Systems Innovation in the Fourth Industrial Revolution Era*, https://doi.org/10.1007/978-3-030-98717-6_1

solutions has thus forced the municipal agencies and the private sector to organize themselves better, and in some cases collaborate to lessen the burden of planning, implementing, and monitoring the performance of provincial projects (Aziz & Mohamad, 2020). This has helped to achieve synergy in infrastructure provision and service delivery. One particular flag project is the Gautrain, which is supported by a shuttle bus system, the Gaubus. The Gautrain is a provincial, state of the art, rail rapid transit system; this is a first, not only in the country but the African continent. At metropolitan municipal levels, the Gauteng province has also witnessed massive innovations in public transport provisions, with the Johannesburg municipality introducing the Rea Vaya public Bus Rapid Transit system and the Tshwane municipality also recently introducing the A Re Yeng public BRT system as a way of easing the pressure on residents as they go about their work, business and leisure trips (Govender, 2014; 2016). GPS data, as well as topographical base maps which show road networks, suburbs, routes of the Gautrain, Gaubus, Rea Vaya and A Re Yeng and car parks, indicate the state of integration of modes and land uses or communities and road conditions such as traffic incidents, congestions, and weather (Chakwizira et al., 2019; Chirisa et al., 2019; Scorcia & Munoz-Raskin, 2019). Furthermore, contemporary public transport provision in developing cities has been characterized by numerous challenges. Moreover, there is a dearth of literature about how technology can facilitate the operation of public transportation systems so as to ease the movement of residents in a developing world context. The Fourth Industrial Revolution has allowed for technological integration in public transportation operations globally (Bahamonde-Birke et al., 2018), however, in developing countries, this is still at the initial phases of implementation and use. This work consequently seeks to unpack the application of smart technology within the transportation needs of growing cities in the developing world. This is done by finding answers to a set of questions that include the following:

What are the existing legislative frameworks and structures which govern smart technology usage in public transport systems in developed and developing cities? What is the current smart technology integration in public transport systems in selected Southern African cities? What is the state of electronic integration in innovative public transport? What is the socio-economic impact of smart technology integration in public transport systems on the lives of commuters? What are the implications and lessons learned from the new emerging developments in innovative public transportation?

1.2 Public Transport Concepts and Issues

Public transport planners in developing countries are faced with challenges of meeting the demand for efficient mobility systems (Musakwa & Gumbo 2017). Given how urban areas have grown to become influential hubs attracting economic activities, the flow of traffic has become an essential element affecting productivity levels (Alam & Ahmed 2013). National and local governments have developed policies such as the National Transport Policy White Paper and the 25-Year Integrated Transport Master Plan which seek to govern urban mobility. However, there continue to be challenges in providing inclusive, reliable, and quality public transport systems (Transport, 2017). Literature on the integration of various public transport systems nevertheless reveals some essential merits such as improved service delivery, interconnectivity of places of economic activity, and improved quality of life for citizens (Mishra, 2019; Ojo, 2019; Scorcia & Munoz-Raskin, 2019). However, the scarcity of longitudinal data impedes the development of a framework that analyses longitudinal and cross-sectional data to analyze the statistical associations between transportation infrastructure and mobility patterns (Huang et al., 2019).

To cope with such constraints, researchers have adopted theoretical frameworks which unpack commuter travel behavior while also focusing on integrating the commuting experience by integrating (a) mobile payments, (Brakewood et al., 2014; Di Pietro et al., 2015; Fontes et al., 2017); (b) route planning (Chowdhury & Ceder, 2016; Jayasinghe et al., 2017); (c) and web 2.0 platforms (Hasan & Ukkusuri, 2014, Moyo & Musakwa, 2016b). Although previous studies conclude that public transportation has the potential to reduce private vehicle usage, there is limited research demonstrating ridership change patterns. From the works of modern-day scholars such as (Filippi et al., 2013), and (Nunes et al., 2014), demand for efficient mobility systems has been a driving force for innovation in transportation planning. In this book, we present how innovations from the Fourth Industrial Revolution (4IR) can alleviate the multifaceted challenges faced in designing and managing public transport systems.

Globally innovative urban public transport systems are geared toward the achievement of the principles of efficiency which include livability, safety, reliability, viability, convenience, accessibility, and cost-effectiveness (Toth-Szabo & Várhelyi, 2012). These principles are used as indicators to monitor sustainability of transport planning and development. Such comprehensive outcomes of sustainable urban mobility should be developed through good governance initiatives in partnership with stakeholder coordination. In defining government and governance, Kobler (2009) referred to government as provision of services to the people, while good governance is the ability to lead a society with differing interests and getting them to work toward a common goal. Considering Kobler's definition of the difference between government and good governance it is essential that there is buy-in of all stakeholders and that everyone is working toward a common goal.

Furthermore, Mangiaracina et al. (2017) have drawn a clear distinction between stakeholder engagement through public participation, and consultation of already formulated development proposals. Thus, improved citizen participation is central to the inclusivity state in policy and plan formulation as well as decision making. All stakeholders affected by the implementation of the envisaged development programs either as investors, project funders, or beneficiaries should consequently form part of every decision-making process such that everyone can take full ownership for ensuring the success of the development program (Baptista, 2005). These interactions must therefore be affected at all levels with all stakeholders being given an opportunity to make meaningful contributions.

Despite the above arguments, the need for stakeholder cooperation has been identified as a crucial component that is lacking, notwithstanding continued massive investments for public transport infrastructure and services by government (Pretorius & Schurink, 2007). Therefore, a need exists to strengthen the public sector human resource with carefully analyzed and understood public transport market trends in order to influence the country toward a comprehensive urban public transport network and implementation of policy and legislation.

Major constraints identified in developing countries by researchers have been the inability of transport and road plans to accommodate buses, pedestrians, and cycling (Ndwandwe & Gumbo, 2020). This has taken place parallel to poor enforcement of legislation. As a result, congestion, problems of livability and all types of pollution have become noticeable, while transport fares always increase.

Social and economic transformation has been arguably the core of South Africa's post-apartheid economic growth and development agenda (Leibbrandt et al., 2012). This has been in response to spatial structures and forms inherited from the apartheid planning system which have hindered and frustrated social and economic transformation in most parts of the country (Seeliger & Turok, 2013). As a consequence, limited progress in the country's implementation of urban public transport planning and socio-economic transformation policies leaves much to be desired, despite good policy instruments and legislation. The main defining feature has been the issue of economic disparities and labor market polarization which persist and coexist with the previously disadvantaged continuously being marginalized.

1.3 Travel Behavior and Smart Mobility

Urban mobility research investigates the dynamics of mode choice, route choice, and travel time in order to model travel behavior. Contemporary travel behavior is changing due to emerging technologies resulting from the Fourth industrial revolution. To cope with these changes, current urban transportation systems need to be reshaped to better support future urban mobility as well as sustainable development. In interpreting travel behavior, a notable study by Nunes et al. (2014) highlighted the importance of co-creation in the era of 4IR by using a prototype mobile application that enabled the exchange of information between commuters and service providers (Assaad et al., 2020). The study proposed that commuters would be willing to share information about their commuting experiences if a platform was made available. This strategy provided a win–win situation between the service provider and the commuter, as information was shared back-and-forth in the co-creation of the public transit system. On the commencement of their trip, commuters would require information on bus travel times, while the service provider required information about the quality of service, based on the commuters' travel experiences. Both parties could greatly benefit from a platform that enabled this interaction. The prototype application utilized spatio-location data to record commuter travel patterns and travel intentions (Mishra et al., 2012). A commuter had the option to check-in once the trip commenced and was given access to communicate with other commuters and also the service provider. Findings from the study revealed that commuters were willing to use the application, if it were available on all platforms, as it would enable them to better plan their trips. Also, service providers benefited from the active commuter engagement in co-creation of the application, as the feedback enables them to obtain insight into the commuters' travel needs. However, the authors also highlighted the importance of having mechanisms to analyze and validate this data so that it could be beneficial to both the commuter and the service provider.

In the past decade, various scholars have explored how co-creation through crowdsourcing can be used to build smarter cities (Marzano et al., 2009). Although the private sector initially championed the development of smart cities, municipalities have also come on board over the years, as the realization of the merits of co-creation promises great rewards for the city at large (Kruger et al., 2018; Ramaswamy & Ozcan, 2018). Currently, within the Southern African context, there has been very little utilization of co-creation in transport planning. This is a critical challenge within the planning domain, as currently there are only a few existing frameworks developed in partnership with ordinary citizens. This insight highlights a knowledge gap, in that planners are not well equipped to deal with contemporary mobility challenges.

In the South African context, some of the causes of having a few co-created frameworks could be due to most planners only utilizing e-planning at novice to intermediate levels (Bajmócy & Pataki, 2019). It is of course inaccurate to say that planners lack the skill to deal with the aforementioned problems, but rather there is a lack of tools to assist planners to address these challenges; hence modern-day planners find themselves caught in a dilemma. Thus, there is a need to develop an enabling platform that merges how planners envision the city's development while addressing cultural, social, and environmental concerns (Boulange et al., 2017). Possible problems of crowdsourcing have been highlighted, as reaching a consensus can be time-consuming. However, one of the merits of using such a platform is that the current understanding and knowledge of stakeholders are taken into consideration, hence modeling the output based on these specifications. Another noteworthy aspect of co-creation during the era of the 4IR is its interactive capability; this enables planners to develop robust regional and local plans, bringing together local and international alliances, including scholars, practitioners, and students in the planning and geo-spatial domains.

Opportunities that exist in co-creation usage in Southern Africa lie in the concept of the smart city and 4IR as points of origin in developing African cities. A strong component of such frameworks is the utilization of e-learning, as this has become increasingly attractive to both academics and community members. This is mainly due to the adaptive nature of e-learning, as it allows the user to interrogate the data while ensuring that co-creation is maintained. (Boulange et al., 2017) have expressed concerns with regard to reconciliation of development trends and sustainability, while de Freitas Miranda and da Silva (2012) advocate for a convergence between what is available and stakeholders. This presents an opportunity to educate stakeholders at all levels about the capabilities of co-creation through crowdsourcing and how these

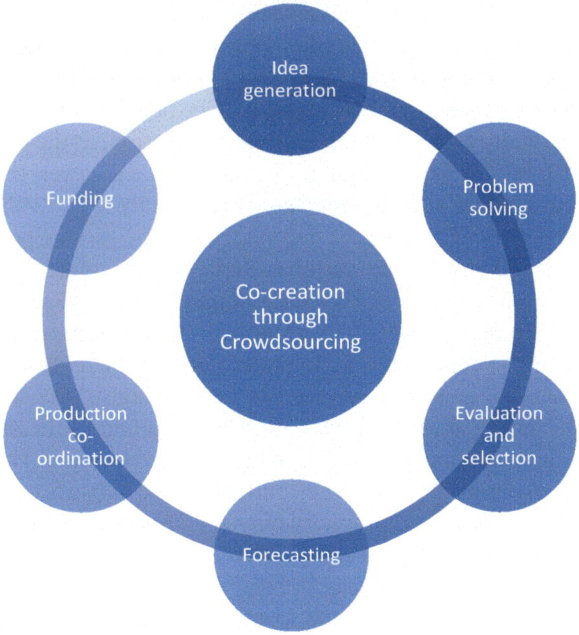

Fig. 1.1 Examples of co-creation through crowdsourcing

tools can be used at a local and national level to promote the development of urban spaces. Notable merits of utilizing crowdsourcing are shown in Fig. 1.1.

Ferreira et al. (2017), in an empirical study, developed an application called OneRide. This gives the commuter the option to use one payment system at various check-in and check-out points of multiple commuting modes (Yatskiv & Budilovich, 2017). The fares are based on distance traveled and the service provider. The route planner incorporates data from the service providers' bus/train schedules with real-time data on departures, arrivals, and available route choices for various trips (Ferreira et al., 2017). The last component is social media, which enables commuters to share information in real-time about their current trip. Findings from the study revealed that commuters were very receptive to the integrated payment system, while service providers also benefited from a significate increase in the number of trips commuters made across the modes. Systems such as innovative public transportation in Gauteng, South Africa require individuals and groups of stakeholders to be socially included in the development process to be sustainable. As the Gautrain has been identified as the backbone of the future public transport system in Gauteng, perhaps the integration of services offered by the Gautrain and the newly implemented Bus Rapid Transport systems can lead to more multi-mode trips (Emeran et al., 2013). Mishra et al. (2012) have also outlined the use of multi-modes as a means of addressing complications relating to mobility in urban areas through promoting accessible ingress and egress of stops or stations. Hence the key focus should be promotion of mobility and connectivity within numerous public transportation providers.

Furthermore, drawing from the rapid growth in the exchange of information, web 2.0 platforms have led to a new phenomenon of big data, which over time has become pervasive (Zhang et al., 2011; Žunić et al., 2020). Yet, its definition is still unclear, as it has been used to express various data types, social media analytics, large quantities of statistics to real-time data. Furthermore, scholars have outlined how many organizations still fail to realize the full potential of big data being produced over the internet by sensors and devices and its use as a tool for decision making (Forsyth & Chitor, 2012). Given the characteristics of big data being namely velocity, variety, and volume, the mining, and analysis of web 2.0 promises an unprecedented data source for transport planning. Now, more specific and interesting is the Southern African case, as most citizens have embraced web 2.0 in their daily lives through Facebook and Twitter (Gavaza et al., 2019; Heyns & Buckley, 2019). Thus, for the cities in the developing world to reach their potential, the development of frameworks and structures is essential. This will guide the growth and integration of existing and future public transport systems while also incorporating the available resources from emerging technologies.

1.4 Methodological Orientation

The study of urban areas involves an orderly investigation of the character and outcomes of political, economic, social environmental, and technological processes which take place at different spatial levels (Griffin & Jiao, 2019; Marchetti & Wanke, 2019; Milne & Watling, 2019). This study, therefore, adopted a case study survey design that applied mixed-method approaches. Research knowledge is derived from data interpretation from the whole to the part and back to the whole (Bonvillain, 2019; Malmberg, 2019).

(a) Policy, legislative, and governance frameworks in urban public transport systems

Document analysis was applied to collect the information which was used to unpack policy, legislative, and governance frameworks in urban public transport systems. Document analysis is a systematic procedure for reviewing or evaluating both printed and electronic records (computer-based and Internet-transmitted). Like other analytical methods in qualitative research, document analysis requires that data be examined and interpreted so as to elicit meaning, gain understanding, and develop empirical knowledge (Corbin & Strauss, 2008). The authors reviewed relevant and related literature on transportation planning and transformation of cities across the country. The literature was undertaken at international, national, and local levels. The search engines that were used included online Elsevier e-library, JSTOR, Google Scholar, Scopus, and SAGE Journals.

The literature review examined development plans and legislation obtained from national and decentralized institutions such as the City of Johannesburg Municipal planning documents, policies, and legislative framework. The documents that were reviewed included the SPLUMA, SDF, GDS, IDP, and Precinct Plans. The review was important to establish the spatial transformation policies and legislative framework potential for improving or constraining urban development in South Africa. Additionally, some structural plans within the city of Johannesburg's jurisdiction were also analyzed to provide insight into the effectiveness of transportation in guiding urban mobility and management.

(b) Physical planning and integration of urban public transport systems

The book utilizes information available from the abovementioned big data sets, and correlation techniques help to assess how proximity affects the commuters' mobility trends while also showing a relation to the availability of infrastructure. To assess the spatial nature of the innovative public transport systems, Geographic Information Systems (GIS) have been used and recommendations are developed to inform the future planning of routes for all public transport modes. These will improve the integration of not only modes but also communities so that they benefit socially and economically (Shen & Stopher, 2013). Through using Arc GIS spatial analyst tools, neglected sections of the province appear clearly on GIS, hence indicating future action plans. For example, the catchment areas and sphere of influence, as well as threshold for businesses, have been determined using neighborhood analysis, buffering, density maps which show hot and cold spots of activity.

As such, a mixed-method approach has been used to gather travel time data, trip planning data, payment methods data, GeoWeb 2.0 data, and GTFS data. Data analysis and interpretations have used content analysis, statistical analysis, and thematic analysis. The study has relied on crowdsourced data administered to commuters within the four metropolitan cities. The questionnaire has been administered online through Google forms and web 2.0 platforms namely Twitter and Facebook, to obtain a good response rate. The questionnaire has both closed and open-ended questions and is centered on four themes outlined in Table 1.1.

Besides the questionnaire, secondary data such as public transportation service providers' reports and literature have been utilized to supplement information gleaned from the questionnaire. Google sheets have helped to analyze the quantitative results of the crowdsourced data to derive descriptive statistics. The query builder functionality (text search, and word count) of Nvivo was utilized to gain better understanding of the responses.

(c) Functional and systematic integration of public transport systems

The connectivity analysis has been computed using the centrality extension in Matlab. Defined by $d(i, j)$, the distance between vertices i and j and the closeness of a bus stop i have been computed as $C_c(i) = \frac{n-1}{\sum_{j \in \subseteq} d(i,j)}$, with n being the number of buses stops in the public transportation network. In this work, the distance has been computed according to the Euclidean distance between bus stops. It should be noted that, though it is highly efficient in determining the influential bus stops, the closeness centrality due to the computational complexity involved in the calculations is not easily applicable in large-scale networks (Brandes et al., 2016).

Table 1.1 Questionnaire themes

Variable	Sub-indicators	Response
Intelligent traffic systems	Automated and electronic ticket system	Yes/No
	Availability of route schedules and waiting times	Yes/No
	Payment options	Yes/No
Public transport uptake	Network density	Quantity in m^2
	Demand for public transit	Yes/No
	Number of commuters	Quantity
	Level of integration	Yes/No
	Robustness of network	Yes/No
Industry 4.0	Route network data availability GIS education	Either open source/not
	Are the public transportation providers adept and ready for Industry 4.0? How far are the public transportation providers in digitalization of operations	Yes/No Rated between 0 and 100% 0 = no digitization 100 = complete digitization

The local Moran's I analysis (Anselin, 1995) has been developed using the centrality indicator, and a heat map has been created for the spatial concentration distribution of nodes. Arc GIS Pro has been used to calculate the local Moran I index to visualize the distribution of nodes along with the public transportation networks. Arc GIS Pro has been chosen as it provides many spatial statistical analysis tools for mapping clusters. To identify the distribution of hot spots along with the public transportation networks and rank these hot spots, the degree centrality has taken account of the network relationship of the bus stops. Consequently, this has been used to articulate why some bus stops were hot spots and others were not. When calculating the local Moran's I index, we have used a fixed distance band to conceptualize the spatial relationships, ensuring all nodes have a neighbor and ensuring a uniform distribution.

$$I_i = \frac{c_i - \overline{C}}{S_i^2}, I \sum_{j=1, j \neq i}^{n} w_{i,j}(c_j - \overline{C}) \tag{1.1}$$

where c_i is the closeness centrality of the bus stop at location i, $\overline{X}$ is the mean of the corresponding attribute, and $w_{i,j}$ is the spatial weight, being the travel time between bus stops i and j:

where n equates to the total number of bus stops.

$$S_i^2 = \frac{\sum_{j=1 j \neq i}^{n} \left(c_j - \overline{C}\right)^2}{n-1} \tag{1.2}$$

Table 1.2 GeoWeb 2.0 dataset for each city

Column ID	Column name	Unit	Interpretation
1	Lat	Degree	Latitude reference of GeoWeb 2.0
2	Lon	Degree	Longitude reference of GeoWeb 2.0
3	Message		Information conveyed in social media post
4	Location		Suburb or city post was made from
5	Source	T or F	Social media platform Twitter (T) or Facebook (F)
6	Date	YYYYMMDD	The day of the month or year
7	Time	HMS	Time of day

Interactive social platforms such as Twitter and Facebook have been used to gather data about current users of public transport showing their points of interest when traversing using the public transportation systems in the cities. This has helped to show the penetration of these public transport platforms within the Gauteng province. This penetration has shown spatial integration and presence of public transport within the province. Travel behaviors assessed reveal the social and economic effects of public transport that is currently being provided in the province. Table 1.2 shows the GeoWeb 2.0 data sets collected for the six cities over a period of 1 year. GeoWeb 2.0 data assists with a more qualitative and quantitative understanding and a deeper insight through language usage, the psychological drivers behind conversations, language and linguistic insight, advanced word clouds, topics, category analysis, mentions and feeds, crucial data in determining commuter behavior and reasons for commuting (Oliveira & Welch, 2013). Strava Metro data have been used to investigate the state of integration between the cycling routes and other modes such as roads and railway tracks.

The collaborative planning framework has been suited in this study as it has informed the implementation and management approaches of innovative urban public transport systems. The framework has helped in providing and ensuring coordinated urban public transport innovations at both national and provincial levels. The frameworks assist in encouraging all relevant stakeholders to work together and should shun working in silos if the goal of providing integrated and well-functioning public transport is to be realized.

1.5 Case Study Areas—Cities in South Africa and Zimbabwe

Southern African cities, particularly the three metropolitan cities of Johannesburg, Tshwane, and Ekurhuleni have improved their public transport systems to include roads, railway, and cycling modes of transport since the beginning of the new millennium. What is not understood well is the state of spatial and communications integration within and among the various modes of transport as well as their impacts on the lives of commuters within the province. Using geographical information systems, big data generated through the use of geo-locational technologies and crowdsourced data from social media networks, this study has investigated the state of smart technology integration in public transport systems across six cities in two developing countries. Preliminary findings have revealed that the novel urban public transport infrastructures and support systems share something. However, a lot still needs to be done to improve their integration and their efficacy toward meeting commuters' needs in terms of connectivity, reliability, convenience, and affordability. Recognizing that developing cities are different, we have assessed innovative public transportation systems in six cities namely Johannesburg, Pretoria, Ekurhuleni, Cape Town, Harare, and Bulawayo (Fig. 1.2). All the cities are run by an elected executive mayor from a political party. The Gauteng economic hub is made up of three metropolitan cities namely Johannesburg, Tshwane, and Ekurhuleni. The innovative public transportation connecting the Gauteng economic hub is the Gautrain, a rapid rail transit system. The Metropolitan city of Johannesburg is the commercial capital of South Africa with a population of 5,926,668 people and a surface area of 1645 km^2 (Stats S.A, 2021). The Rea Vaya is the rapid bus transit system that connects commuters from the south of the city to the central business district of Johannesburg. The Metropolitan city of Pretoria is located to the north of Johannesburg and is the administrative capital

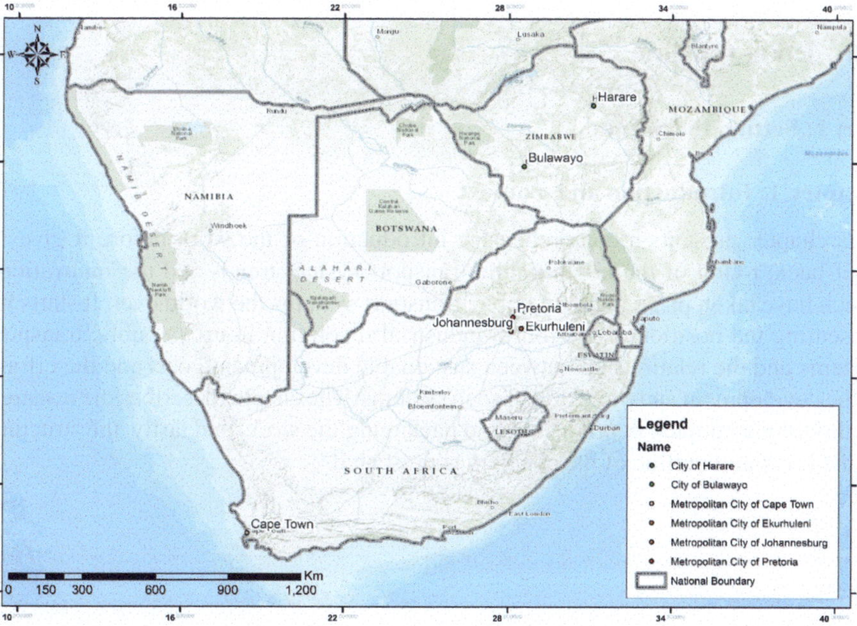

Fig. 1.2 Study area

of South Africa. It has a surface area of 687.5 km^2 and a population of 2,655,000 people (Stats S.A, 2021). The A Re Yeng is a rapid bus transit system that connects commuters within central Pretoria.

The Metropolitan city of Ekurhuleni lies to the east of Johannesburg and is known as South Africa's manufacturing hub with a population of nearly 3,970,000 and a surface area of 1975 km^2 (Stats S.A, 2021). The Harambe is a rapid bus transit system currently being implemented in Ekurhuleni, but it is still in its initial stages of operation. The Metropolitan city of Cape Town, located in the Western Cape province of South Africa has a population of 3,433,441 people and a surface area of 400.3 km^2 (Stats S.A, 2021). The MyCiti rapid bus transit system connects commuters in central Cape Town. Harare is Zimbabwe's, biggest and capital city with a surface area of 960 km^2 and a population of 1,542,000 people. Harare is located in central Zimbabwe whereas Bulawayo is located in the southwest. Bulawayo is Zimbabwe's second-biggest city with a surface area of 1706 km^2 and a population of 1.2 million. Bulawayo has been known as the country's industrial hub, but industry has been declining of late due to the cycles of economic decline in Zimbabwe. All of the six cities are run by an elected executive mayor from a political party. Authorities have recommended that, for the cities to become functional, their economic growth should be supported by an efficient system of public transportation to address urban mobility challenges. The development and monitoring of integrated land-use and transportation planning systems have the potential to inform robust decision-making in developing cities, thus enhancing efficiency.

1.6 Book Outline

Part 1: Setting the Scene

Chapter 1: Introduction and Context

This chapter presents a comprehensive introduction of the work before it gives a brief background of the urban public transport system trends and the innovations which have taken place in urban public transport systems the world over. It starts by presenting the position of the fourth industrial revolution in urban public transport systems and the relationship between sustainable development goals and the efforts in the development and provision of such systems. The chapter describes the research methodology adopted in producing and packaging the work, and lastly, the structure of the book and outline of the chapters is presented.

Part 2: Policy, Legislative and Governance Frameworks in Urban Public Transport Systems in South Africa and Zimbabwe

Chapter 2: Governance and Structures in Urban Public Transport in South African and Zimbabwean Cities

This chapter describes what constitutes good policy, legislative, and governance frameworks in national and international Urban Public Transportation systems. The chapter combines bibliometric analysis of literature and case studies to understand the elements of good policy, and legislative and governance frameworks. Consequently, the chapter identifies the lessons learned from international experiences to aid in the creation of solutions to produce good governance in urban public transportation.

Chapter 3: Role of Policy and Legislative Instruments in South African and Zimbabwean cities

This chapter builds on lessons learned from the previous chapter by exploring and deliberating on the role of current policy and legislative instruments to facilitate the growth of innovative public transportation systems in the era of the Fourth Industrial Revolution in cities in the developing world. To unpack this role, we use a systematic method combining bibliometric and literature review analysis. Findings reveal a lack of a long-term implementation horizon coupled with a lack of stakeholder cooperation and integration within and across government spheres. Policy and legislative frameworks, although they are in support of innovative public transportation, fail to be all-inclusive as they currently only play a supporting role, with most innovative initiatives originating from the private sector.

Part 3: Physical Planning and Integration of Urban Public Transport Systems in South African and Zimbabwean Cities

Chapter 4: Internet of Things and Urban Public Transport Systems in Johannesburg, Tshwane, Ekurhuleni, Cape Town, and Bulawayo

The chapter highlights the internet of things and how this contributes to improving public transportation systems in the developing world. The researchers have focused on case studies of urban public transportation systems, and findings reveal that emerging technologies have been introduced in public transportation. For emerging public transportation systems such as the BRTs and RRTs to achieve the benefits of the internet of things, more investment is required to ensure they are woven holistically into the transport operations and integrated with other public transportation systems.

Chapter 5: Integration of Urban Mobility Systems in the Gauteng City Region

In this chapter, we have used efficient, societal, ecological, and economic constraints to measure connectivity levels at interchange zones. Consequently, the analysis has used indicators to determine variations of connectivity levels across the Gauteng City Region. The chapter reveals hot spots of high interchange zones at major mobility hubs. Results demonstrate that improving connectivity between various modes of public transportation will lead to an increased ridership of public transportation.

Part 4: Functional and Systematic Integration of Public Transport Systems in South Africa and Zimbabwe

Chapter 6: Functional Integration of Public Transport Systems in Tshwane, Johannesburg, Harare, and Bulawayo Cities

This chapter evaluates functional integration (institutionally and operationally) in cities in South Africa and Zimbabwe. Mixed methods such as questionnaire administration, crowdsourced cognitive data, geo-spatial data and semi-structured interviews for data collection with analysis have been followed by thematic analysis. Although functionally disintegrated, at least the billing systems from the BRT systems across the three metropolitan cities are integrated, while the Gautrain is completely disintegrated from the BRT system. Evidently, this has compromised the attainment of Smart Mobility principles of efficiency, accessibility, convenience, reliability, and affordability through the implementation of innovative urban public transport systems. Subsequently, this chapter recommends institutional and operational integration, where route planning, scheduling, route management, and billing systems have been managed through a single-window coordination approach to enhance attainment of Smart Mobility Principles.

Chapter 7: Emerging Developments in Information Communication Technology and Commuting Trends in Johannesburg, Tshwane, Ekurhuleni, and Cape Town Cities

The chapter highlights the incorporation of 4IR in public transportation systems within a Southern African context. Interesting applications in 4IR solutions are notable in urban public transport systems such as Gautrain but are skewed in other modes such as BRTs (Rea Vaya, Are Yeng, and Harambe). Applications have been developed in notable urban public transport systems such as Gautrain. These applications have the capacity to capture commuter movement patterns with minimal costs and in real-time. Given the limited technological innovations in the transport sector, perhaps smart technologies' solutions such as smart stations, smart ticketing, and biometrics solutions when applied can enable real-time data mining for traveling information for all public transportation modes. Therefore, the chapter recommends that systems be integrated electronically. This will create seamless traveling and more efficiency in urban public transport networks using one e-smart card and information scheduling connected with different systems allowing smooth switch in-betweens. In conclusion, considering the limited technological innovations in the transport sector, it is thus hoped that smart technology solutions such as smart stations, smart ticketing and biometrics solutions are applied to enable real-time data for traveling information.

Part 5: Sustainability Considerations in Urban Public Transport Systems in South Africa and Zimbabwe

Chapter 8: Measuring Economic Benefits of Innovative Urban Public Transportation in the Era of 4IR Tshwane, Ekurhuleni and Bulawayo Cities

The chapter highlights the minimal concentrated and meaningful effort which has been made to create economic benefits in the planning and implementation of public transport infrastructure initiatives. The lack of collective strategy formulation by public institutions responsible for public transport infrastructure rollout and those responsible for economic growth and development leaves much to be desired. Evidently, fostering economic benefits as part of public transport infrastructure investments still needs to be incorporated as a crucial component in the planning and implementation of public transport infrastructure projects. A transversal management approach is thus recommended, where interdependent institutions and stakeholders can collaboratively work together to foster economic benefits which form the core of Public Transport Infrastructure Investments initiatives.

Chapter 9: Modeling the Societal Impact of Innovative Urban Public Transport Systems in South African and Zimbabwean Cities

This chapter builds on the previous chapter by exploring the social impact of innovative urban public transport. It highlights the fact that transport institutions have existing policies in place and have established corporate social responsibility functions. However, most existing public transportation providers in the metropolitan cities of Johannesburg, Pretoria, and Ekurhuleni have failed to capture the cognitive elements influencing commuter mobility trends. Thus, the role of citizen-based innovation in public transportation is misplaced.

Part 6: Conclusion

Chapter 10: Innovative Urban Public Transport Systems: An Evolving Continuous Expedition in Southern African Cities

The chapter wraps up the work presented in this book with a comprehensive but concise introduction of the main issues that have emerged from the preceding discussions. It presents the implications and lessons learned from the study as well as makes a case for action by the various stakeholders who derive benefits from integrated urban public transport systems. The chapter also recommends various strategies which, if adopted, could ensure that sustainable and integrated urban public transport systems are provided in other cities of the developing world. Use could be made of the efforts that have been deployed by the metropolitan cities in Gauteng province and the lessons learned from the current case studies. The chapter ends by suggesting new areas for further investigation which could be packaged into another volume. It lastly provides a conclusion to the journey traveled.

References

Alam, M. A., & Ahmed, F. (2013). Urban transport systems and congestion: A case study of Indian cities. *Transport and Communications Bulletin for Asia and the Pacific, 82*, 33–43.

Anselin, L. (1995). Local indicators of spatial association—LISA. *Geographical Analysis, 27*, 93–115.

Arnold, K., Le Roux, A., & Hattingh, M. (2017). Impact of Gautrain stations on property prices and sales activity in the city of Johannesburg between 2006 and 2015. *South African Journal of Geomatics, 6*, 184–195.

Assaad, R., Dagli, C., & El-Adaway, I. H. (2020). A system-of-systems model to simulate the complex emergent behavior of vehicle traffic on an urban transportation infrastructure network. *Procedia Computer Science, 168*, 139–146.

Aziz, A., & Mohamad, J. (2020). Public transport planning: Local bus service integration and improvements in Penang, Malaysia. *Planning Malaysia, 18*.

Bahamonde-Birke, F. J., Kickhöfer, B., Heinrichs, D., & Kuhnimhof, T. (2018). A systemic view on autonomous vehicles: Policy aspects for a sustainable transportation planning. *disP-The Planning Review, 54*, 12–25.

Bajmócy, Z., & Pataki, G. (2019). Responsible research and innovation and the challenge of co-creation.

Baptista, I. (2005). Is there a theory of collaborative planning we can talk about. In *46th AESOP Conference "The Dream of a Greater Europe"*.

Beimborn, E. A., Greenwald, M. J., & Jin, X. (2003). Accessibility, connectivity, and captivity: Impacts on transit choice. *Transportation Research Record*.

Beukes, E., Vanderschuren, M., & Zuidgeest, M. (2011). Context sensitive multimodal road planning: A case study in Cape Town, South Africa. *Journal of Transport Geography, 19*, 452–460.

Bokopane, L., Kusakana, K., & Vermaak, H. J. (2015). Optimal energy management of an isolated electric Tuk-Tuk charging station powered by hybrid renewable systems. In *2015 International Conference on the Domestic Use of Energy (DUE)* (pp. 193–201). IEEE.

Bonvillain, N. (2019). *Language, culture, and communication: The meaning of messages*. Rowman & Littlefield.

Boulange, C., Gunn, L., Giles-Corti, B., Mavoa, S., Pettit, C., & Badland, H. (2017). Examining associations between urban design attributes and transport mode choice for walking, cycling, public transport and private motor vehicle trips. *Journal of Transport & Health, 6*, 155–166.

Brakewood, C., Rojas, F., Robin, J., Sion, J., & Jordan, S. (2014). Forecasting mobile ticketing adoption on commuter rail. *Journal of Public Transportation, 17*, 1.

Brandes, U., Borgatti, S. P., & Freeman, L. C. (2016). Maintaining the duality of closeness and betweenness centrality. *Social Networks, 44*, 153–159.

Ceder, A. (2020). Urban mobility and public transport: Future perspectives and review. *International Journal of Urban Sciences*, 1–25.

Černá, L., Zitrický, V., & Abramović, B. (2020). Methodical manual for a set of transport regulations in railway passenger transport. *LOGI—Scientific Journal on Transport and Logistics, 11*, 13–24.

Chai, C., Fan, J., Li, G., Wang, J., & Zheng, Y. (2019). Crowdsourcing database systems: Overview and challenges. In *2019 IEEE 35th International Conference on Data Engineering (ICDE)* (pp. 2052–2055). IEEE.

Chakwizira, J. (2020). The intersection of spatial fragmentation and smart transport planning in Gauteng Province, South Africa: Constraints and opportunities. In *International Conference on Applied Human Factors and Ergonomics* (pp. 367–373). Springer.

Chakwizira, J., Bikam, P., & Adeboyejo, T. A. (2019). Transport realities and challenges for low income peripheral located settlements in Gauteng Province: Are we witnessing the genesis of a new transport order or consolidation of the old transport order? In *International Conference on Applied Human Factors and Ergonomics* (pp. 618–630). Springer.

Chakwizira, J., Bikam, P., & Adeboyejo, T. A. (2021). Access and constraints to commuting for persons with disabilities in Gauteng Province, South Africa. In *Urban inclusivity in Southern Africa*. Springer.

Charman, A., Tonkin, C., Denoon-Stevens, S., & Demeestere, R. (2017). *Post-apartheid spatial inequality: Obstacles of land use management on township micro-enterprise formalisation*. A report by the Sustainable Livelihoods Foundation. Sustainable Livelihoods Foundation.

Chirisa, I., Mhlanga, G., & Matamanda, A. R. (2019). Infrastructure, city-region development, and Africa's territorial spaces: Gauteng as the exemplar. In *Optimizing regional development through transformative urbanization*. IGI Global.

Chowdhury, S., & Ceder, A. A. (2016). Users' willingness to ride an integrated public-transport service: A literature review. *Transport Policy, 48*, 183–195.

Chowdhury, S., Hadas, Y., Gonzalez, V. A., & Schot, B. (2018). Public transport users' and policy makers' perceptions of integrated public transport systems. *Transport Policy, 61*, 75–83.

Cramer, J., & Krueger, A. B. (2016). Disruptive change in the taxi business: The case of Uber. *American Economic Review, 106*(5), 177–182.

Corbin, J., & Strauss, A. (2008). Basics of Qualitative Research: Techniques and Procedures for Developing Grounded Theory (3rd ed.). Thousand Oaks, CA: Sage.

Culwick, C., Gotz, G., Katumba, S., Trangoš, G., & Wray, C. (2015). Mobility patterns in the Gauteng city-region, South Africa. *Regional Studies, Regional Science, 2*, 309–311.

Dawood, G., & Mokonyama, M. (2015). Towards a more optimal passenger transport system for South Africa: Design of public transport operating subsidies. *Council for Scientific and Industrial Research (CSIR), 62*, 192–213.

de Freitas Miranda, H., & da Silva, A. N. R. (2012). Benchmarking sustainable urban mobility: The case of Curitiba, Brazil. *Transport Policy, 21*, 141–151.

Department of Transport. (2017). Draft Revised White Paper on National Transport Policy.

Di Pasquale, G., dos Santos, A. S., Leal, A. G., & Tozzi, M. (2016). Innovative public transport in Europe, Asia and Latin America: A survey of recent implementations. *Transportation Research Procedia, 14*, 3284–3293.

Di Pietro, L., Mugion, R. G., Mattia, G., Renzi, M., & Toni, M. (2015). The integrated model on mobile payment acceptance (IMMPA): An empirical application to public transport. *Transportation Research Part C: Emerging Technologies, 56*, 463–479.

Dubina, I. N., Campbell, D. F., Carayannis, E. G., Chub, A. A., Grigoroudis, E., & Kozhevina, O. V. (2017). The balanced development of the spatial innovation and entrepreneurial ecosystem based on principles of the systems compromise: A conceptual framework. *Journal of the Knowledge Economy, 8*, 438–455.

Dumba, S. (2017). Informal public transport driver behaviour and regulatory policy linkage: An expose. *Journal of Transport and Supply Chain Management, 11*(1), 1–16.

Eggleston, H., Buendia, L., Miwa, K., Ngara, T., & Tanabe, K. (2006). 2006 IPCC guidelines for national greenhouse gas inventories.

Emeran, H., Heyns, W., & Dyer, R. (2013). PRASA's new National Strategic Plan: transportation engineering. *Civil Engineering= Siviele Ingenieurswese, 2013*, 12–15.

Ferreira, M. C., Fontesz, T., Costa, V., Dias, T. G., Borges, J. L., & de Cunha, J. F. (2017). Evaluation of an integrated mobile payment, route planner and social network solution for public transport. *Transportation Research Procedia, 24*, 189–196.

Filippi, F., Fusco, G., & Nanni, U. (2013). User empowerment and advanced public transport solutions. *Procedia-Social and Behavioral Sciences, 87*, 3–17.

Fombad, M. C. (2015). Governance in public–private partnerships in South Africa: Some lessons from the Gautrain. *Journal of Southern African Studies, 41*, 1199–1217.

Fontes, T., Costa, V., Ferreira, M. C., Shengxiao, L., Zhao, P., & Dias, T. G. (2017). Mobile payments adoption in public transport. *Transportation Research Procedia, 24*, 410–417.

Forsyth, C., & Chitor, R. (2012). *For big data analytics there's no such thing as too big: The compelling economics and technology of big data computing*. White Paper. Forsyth Communications.

Gabdrakhmanova, N., & Pilgun, M. (2021). Cognitive systems, artificial neural networks and differential equations: Social media data. *Procedia Computer Science, 186*, 677–684.

Gauthier, A., & Weinstock, A. (2010). Africa transforming paratransit into BRT. *Built Environment, 36*, 317–327.

Gavaza, B. K., Viljoen, K. L., & Cilliers, L. (2019). The influence of social media service quality on client loyalty in the South African banking industry. *Acta Commercii, 19*, 1–10.

Ge, M., Bangui, H., & Buhnova, B. (2018). Big data for internet of things: A survey. *Future Generation Computer Systems, 87*, 601–614.

Gössling, S., & Choi, A. S. (2015). Transport transitions in Copenhagen: Comparing the cost of cars and bicycles. *Ecological Economics, 113*, 106–113.

Gotz, G., Wray, C., Venter, C., Badenhorst, W., Trangoš, G., & Culwick, C. (2014). Mobility in the Gauteng city-region.

Govender, K. K. (2014). Public transport service quality in South Africa: A case study of bus and mini bus services in Johannesburg. *African Journal of Business Management, 8*, 317–326.

Govender, K. K. (2016). Exploring public transport service quality: The case of mini-bus taxi service in South Africa. *Eurasian Business Review, 6*, 101–116.

Govender, R., & Allopi, D. (2006). Towards a safer minibus taxi industry in South Africa. In *SATC 2006*.

Griffin, G. P., & Jiao, J. (2019). Crowdsourcing bicycle volumes: Exploring the role of volunteered geographic information and established monitoring methods.

Gumbo, T., Ingwani, E., Mufungizi, A. A., & Dumba, S. (2021). Planning for climate and cyclone resilient transport infrastructures in Southern Africa: Lessons from Zimbabwe. In *Cyclones in Southern Africa*. Springer.

Gumbo, T., & Moyo, T. (2020). Exploring the interoperability of public transport systems for sustainable mobility in developing cities: Lessons from Johannesburg Metropolitan City, South Africa. *Sustainability, 12*, 5875.

Guzman, L. A., Arellana, J., & Alvarez, V. (2020). Confronting congestion in urban areas: Developing sustainable mobility plans for public and private organizations in Bogotá. *Transportation Research Part A: Policy and Practice, 134*, 321–335.

Gwala, S. (2007). Urban non-motorised transport (NMT): A critical look at the development of Urban NMT Policy and planning mechanisms in South Africa from 1996–2006. In *SATC 2007*.

Hasan, S., & Ukkusuri, S. V. (2014). Urban activity pattern classification using topic models from online geo-location data. *Transportation Research Part c: Emerging Technologies, 44*, 363–381.

Henama, U. S., & Sifolo, P. P. S. (2017). Uber: The South Africa experience. *African Journal of Hospitality, Tourism and Leisure, 6*(2), 1–10.

Heyns, W., & Buckley, S. A. (2019). Twitter knowledge sharing model based on small businesses in the Western Cape. In *2019 Conference on Information Communications Technology and Society (ICTAS)* (pp. 1–5). IEEE.

Hill, N., Gibson, G., Guidorzi, E., Amaral, S., Parlikad, A., & Jin, Y. (2017). Scoping study into deriving transport benefits from big data and the internet of things in smart cities.

Hook, W., & Howe, J. (2005). *A background paper to the task force on slum dwellers of the millennium project*. ITDP.

Horn, A. (2020). The darker side of muddling through: An analysis of spatial planning policy decision-making and policy termination in Cape Town, South Africa. In *Urban Forum* (pp. 1–26). Springer.

Hu, X., Chang, S., Li, J., & Qin, Y. (2010). Energy for sustainable road transportation in China: Challenges, initiatives and policy implications. *Energy, 35*, 4289–4301.

Huang, Q., & Wong, D. W. (2016). Activity patterns, socioeconomic status and urban spatial structure: What can social media data tell us? *International Journal of Geographical Information Science, 30*, 1873–1898.

Huang, X., Cao, X. J., Yin, J., & Cao, X. (2019). Can metro transit reduce driving? Evidence from Xi'an, China. *Transport Policy, 81*, 350–359.

Hussein, W. N., Kamarudin, L., Hussain, H. N., Zakaria, A., Ahmed, R. B., & Zahri, N. (2018). The prospect of internet of things and big data analytics in transportation system. In *Journal of Physics: Conference Series* (p. 012013).
Jayasinghe, A., Sano, K., & Rattanaporn, K. (2017). Application for developing countries: Estimating trip attraction in urban zones based on centrality. *Journal of Traffic and Transportation Engineering (english Edition), 4*, 464–476.
Jennings, G. (2015). A bicycling renaissance in South Africa? Policies, programmes & trends in Cape Town. In *Southern African Transport Conference.*
Jiang, Y., Li, Z., & Ye, X. (2019). Understanding demographic and socioeconomic biases of geotagged Twitter users at the county level. *Cartography and Geographic Information Science, 46*, 228–242.
Kemajou, A., Jaligot, R., Bosch, M., & Chenal, J. (2019). Assessing motorcycle taxi activity in Cameroon using GPS devices. *Journal of Transport Geography, 79*, 102472.
Kobler, A. M. (2009). *Building community capacity: How collaborative planning is changing the culture of governance in Seattle.* Iowa State University.
Krishankumar, R., Pamucar, D., Deveci, M., & Ravichandran, K. S. (2021). Prioritization of zero-carbon measures for sustainable urban mobility using integrated double hierarchy decision framework and EDAS approach. *Science of The Total Environment, 797*, 149068.
Kruger, C., Caiado, R. G. G., França, S. L. B., & Quelhas, O. L. G. (2018). A holistic model integrating value co-creation methodologies towards the sustainable development. *Journal of Cleaner Production, 191*, 400–416.
Kumar, S., Tiwari, P., & Zymbler, M. (2019). Internet of Things is a revolutionary approach for future technology enhancement: A review. *Journal of Big Data, 6*, 1–21.
Larsen, J. (2017). The making of a pro-cycling city: Social practices and bicycle mobilities. *Environment and Planning A, 49*, 876–892.
Leibbrandt, M., Finn, A., & Woolard, I. (2012). Describing and decomposing post-apartheid income inequality in South Africa. *Development Southern Africa, 29*, 19–34.
Lesh, M. C. (2013). Innovative concepts in first-last mile connections to public transportation. In *Urban public transportation systems 2013.*
Levy Abitbol, J., Fleury, E., & Karsai, M. (2019). Optimal proxy selection for socioeconomic status inference on twitter. *Complexity.*
Li, V. O., Lam, J. C., Han, Y., & Chow, K. (2021). A big data and artificial intelligence framework for smart and personalized air pollution monitoring and health management in Hong Kong. *Environmental Science & Policy, 124*, 441–450.
Lomme, R. (2008). Should South African minibus taxis be scrapped? Formalizing informal urban transport in a developing country. In *Proceedings of the CODATU XIII Conference*, Ho Chi Minh City.
Loubser, J., Marnewick, A. L., & Joseph, N. (2021). Framework for the potential userbase of mobility as a service. *Research in Transportation Business & Management, 39*, 100583.
Lucas, K., & Markovich, J. (2011). International perspectives. In *New perspectives and methods in transport and social exclusion research.*
Malmberg, T. (2019). *Human territoriality: Survey on the behavioural territories in man with preliminary analysis and discussion of meaning.* Walter de Gruyter GmbH & Co KG.
Mangiaracina, R., Perego, A., Salvadori, G., & Tumino, A. (2017). A comprehensive view of intelligent transport systems for urban smart mobility. *International Journal of Logistics Research and Applications, 20*, 39–52.
Marchetti, D., & Wanke, P. F. (2019). Efficiency in rail transport: Evaluation of the main drivers through meta-analysis with resampling. *Transportation Research Part a: Policy and Practice, 120*, 83–100.
Marzano, G., Lizut, J., & Siguencia, L. O. (2019). Crowdsourcing solutions for supporting urban mobility. *Procedia Computer Science, 149*, 542–547.

May, A., Boehler-Baedeker, S., Delgado, L., Durlin, T., Enache, M., & van der Pas, J.-W. (2017). Appropriate national policy frameworks for sustainable urban mobility plans. *European Transport Research Review, 9*, 7.

Mazumdar, S., & Thakker, D. (2020). Citizen science on Twitter: Using data analytics to understand conversations and networks. *Future Internet, 12*, 210.

Mbara, T., & Pisa, N. (2018). An analysis of impediments to deliver sustainable transport in cities of the developing countries: The case of Harare, Zimbabwe.

Mbara, T. C. (2016). Tuk-tuk, 'new kid on the block' in Johannesburg: Operational and user travel characteristics, competition and impacts. *Journal of Transport and Supply Chain Management, 10*(1), 1–9.

Mbatha, S., & Gumbo, T. (2019). Establishing the state of spatial integratedness of innovative public transport systems in Gauteng, South Africa. In *REAL CORP 2019—IS THIS THE REAL WORLD? Perfect Smart Cities vs. Real Emotional Cities. Proceedings of 24th International Conference on Urban Planning, Regional Development and Information Society* (pp. 319–327). CORP–Compentence Center of Urban and Regional Planning.

McKay, T., Simpson, Z., & Patel, N. (2017). Spatial politics and infrastructure development: Analysis of historical transportation data in Gauteng-South Africa (1975–2003). *Miscellanea Geographica. Regional Studies on Development, 21*, 35–43.

Milne, D., & Watling, D. (2019). Big data and understanding change in the context of planning transport systems. *Journal of Transport Geography, 76*, 235–244.

Mishra, A. K. (2019). Cities, transport and agglomeration: Addressing the urban mobility challenges in India. *Growth and Change, 50*, 1115–1133.

Mishra, S., Welch, T. F., & Jha, M. K. (2012). Performance indicators for public transit connectivity in multi-modal transportation networks. *Transportation Research Part a: Policy and Practice, 46*, 1066–1085.

Moradi, A., & Vagnoni, E. (2018). A multi-level perspective analysis of urban mobility system dynamics: What are the future transition pathways? *Technological Forecasting and Social Change, 126*, 231–243.

Moyo, T., Kibangou, A. Y., & Musakwa, W. (2021). Societal context-dependent multi-modal transportation network augmentation in Johannesburg, South Africa. *Plos one, 16*, e0249014.

Moyo, T., & Musakwa, W. (2016a). Using crowdsourced data (Twitter & Facebook) to delineate the origin and destination of commuters of the Gautrain public transit system in South Africa. *ISPRS Annals of Photogrammetry, Remote Sensing & Spatial Information Sciences, 3.*

Moyo, T., & Musakwa, W. (2016b). Using crowdsourced data (Twitter & Facebook) to delineate the origin and destination of commuters of the Gautrain public transit system in South Africa.

Moyo, T., Musakwa, W., & Mokoena, B. (2018). An analysis to investigate spatial cognitive factors which influence cycling patterns in Johannesburg. *International Archives of the Photogrammetry, Remote Sensing & Spatial Information Sciences.*

Musakwa, W., & Gumbo, T. (2017). *Impact of urban policy on public transportation in Gauteng, South Africa: Smart or dumb city systems is the question.* Springer.

Musakwa, W., & Selala, K. M. (2016). Mapping cycling patterns and trends using Strava Metro data in the city of Johannesburg, South Africa. *Data in Brief, 9*, 898–905.

Nakamura, F., Makimura, K., & Toyama, Y. (2017). Perspective on an urban transportation strategy with BRT for developing cities. *Engineering and Applied Science Research, 44*, 196–201.

Ndwandwe, B., & Gumbo, T. (2020). Vibrant or dull urban spaces: Are city of Tshwane "A Re Yeng" and "Gautrain" Connector points places of social and economic interaction? In *Real corp 2020—Shaping urban change livable city regions for the 21st century. Proceedings of 25th International Conference on Urban Planning and Regional Development in the Information Society.*

Nunes, A. A., Galvão, T., & de Cunha, J. F. (2014). Urban public transport service co-creation: Leveraging passenger's knowledge to enhance travel experience. *Procedia-Social and Behavioral Sciences, 111*, 577–585.

Ojo, T. K. (2019). Quality of public transport service: An integrative review and research agenda. *Transportation Letters, 11*, 104–116.

Oliveira, G. H. M., & Welch, E. W. (2013). Social media use in local government: Linkage of technology, task, and organizational context. *Government Information Quarterly, 30*, 397–405.

Ponkshe, A., & Pricing, C. (2020). Policymaking towards green mobility in India. *Occasional Papers.*

Porru, S., Misso, F. E., Pani, F. E., & Repetto, C. (2020). Smart mobility and public transport: Opportunities and challenges in rural and urban areas. *Journal of Traffic and Transportation Engineering (english Edition), 7*, 88–97.

Pretorius, D., & Schurink, W. (2007). Enhancing service delivery in local government: The case of a district municipality. *SA Journal of Human Resource Management, 5*, 19–29.

Pribyl, O., Blokpoel, R., & Matowicki, M. (2020). Addressing EU climate targets: Reducing CO_2 emissions using cooperative and automated vehicles. *Transportation Research Part D: Transport and Environment, 86*, 102437.

Rahman, M.S.-U. (2013). *Integrating BRT with rickshaws in developing cities: A case study on Dhaka City*. University of Leeds.

Ramaswamy, V., & Ozcan, K. (2018). What is co-creation? An interactional creation framework and its implications for value creation. *Journal of Business Research, 84*, 196–205.

Rode, P., Floater, G., Thomopoulos, N., Docherty, J., Schwinger, P., Mahendra, A., & Fang, W. (2017). Accessibility in cities: Transport and urban form. *Disrupting Mobility*, 239–273.

Ryseck, B. (2019). South and Southern Africa—Are disruptive transport technologies ready? An exploration of journey planning tools for urban public transport systems in South Africa. In *Southern African Transport Conference.*

Sanchez-Atondo, A., Garcia, L., Calderon-Ramirez, J., Gutiérrez-Moreno, J. M., & Mungaray-Moctezuma, A. (2020). Understanding public transport ridership in developing countries to promote sustainable urban mobility: A case study of Mexicali, Mexico. *Sustainability, 12*, 3266.

Santos, T., Silva, M. A., Fernandes, V. A., & Marsden, G. (2020). Resilience and vulnerability of public transportation fare systems: The case of the city of Rio De Janeiro, Brazil. *Sustainability, 12*, 647.

Scorcia, H., & Munoz-Raskin, R. (2019). Why South African cities are different? Comparing Johannesburg's Rea Vaya bus rapid transit system with its Latin American siblings. *Case Studies on Transport Policy, 7*, 395–403.

Sebola, M. (2014). Recapitalizing mini-bus taxis for effective public transportation in South Africa: The urban rural transport connection problem. *Urban Transport XX, 138*, 125.

Seeliger, L., & Turok, I. (2013). Towards sustainable cities: Extending resilience with insights from vulnerability and transition theory. *Sustainability, 5*, 2108–2128.

Shen, L., & Stopher, P. R. (2013). A process for trip purpose imputation from global positioning system data. *Transportation Research Part c: Emerging Technologies, 36*, 261–267.

Skarin, F., Olsson, L. E., Friman, M., & Wästlund, E. (2019). Importance of motives, self-efficacy, social support and satisfaction with travel for behavior change during travel intervention programs. *Transportation Research Part f: Traffic Psychology and Behaviour, 62*, 451–458.

Statistics South Africa. (2021). http://www.statssa.gov.za/

Sun, Y., & Cui, Y. (2018). Evaluating the coordinated development of economic, social and environmental benefits of urban public transportation infrastructure: Case study of four Chinese autonomous municipalities. *Transport Policy, 66*, 116–126.

Sun, Y., Du, Y., Wang, Y., & Zhuang, L. (2017). Examining associations of environmental characteristics with recreational cycling behaviour by street-level Strava data. *International Journal of Environmental Research and Public Health, 14*, 644.

Takahashi, T. (2017). Economic analysis of tariff integration in public transport. *Research in Transportation Economics, 66*, 26–35.

Togo, M. (2016). Adoption of green technologies in the public transport sector: The case of Rea Vaya BRT system, South Africa. *Milestones in Green Transition and Climate Compatible Development in Eastern and Southern Africa*, 93.

Toro, E. E. M., van der Krogt, A., & Flores, R. S. (2019). Mobility and integration of public transport systems in Latin America. In *Proceedings of the 2019 2nd International Conference on Machine Learning and Machine Intelligence* (pp. 58–62).

Toth-Szabo, Z., & Várhelyi, A. (2012). Indicator framework for measuring sustainability of transport in the city. *Procedia-Social and Behavioral Sciences, 48*, 2035–2047.

Van Der Westhuizen, J. (2007). Glitz, glamour and the Gautrain: Mega-projects as political symbols. *Politikon, 34*, 333–351.

Van Gerven, M., & Bohte, S. (2017). Artificial neural networks as models of neural information processing. *Frontiers in Computational Neuroscience, 11*, 114.

Van Ryneveld, P. (2008). *15 year review of public transport in South Africa with emphasis on metropolitan areas*. Unpublished paper commissioned for the Fifteen Year Review.

Vascik, P. D., & Hansman, R. J. (2018). Scaling constraints for urban air mobility operations: Air traffic control, ground infrastructure, and noise. In *2018 Aviation Technology, Integration, and Operations Conference* (p. 3849).

Vasconcellos, E. A. (2014). *Urban transport environment and equity: The case for developing countries*. Routledge.

Vecchio, G., & Tricarico, L. (2019). "May the force move you": Roles and actors of information sharing devices in urban mobility. *Cities, 88*, 261–268.

Veeneman, W., & Mulley, C. (2018). Multi-level governance in public transport: Governmental layering and its influence on public transport service solutions. *Research in Transportation Economics, 69*, 430–437.

Venter, M., Steynberg, L., Grundling, J. P., & Feng, X. (2019). Perceived contribution of township enterprises on local economic development in Mabopane Township, South Africa. *Journal of Public Administration, 54*, 888–907.

Vermesan, O., Friess, P., Guillemin, P., Sundmaeker, H., Eisenhauer, M., Moessner, K., Le Gall, F., & Cousin, P. (2013). Internet of things strategic research and innovation agenda. In *Internet of things: Converging technologies for smart environments and integrated ecosystems*. River Publishers.

Vermicelli, S., Cricelli, L., & Grimaldi, M. (2021). How can crowdsourcing help tackle the COVID-19 pandemic? An explorative overview of innovative collaborative practices. *R&D Management, 51*, 183–194.

Walters, J. (2013). Overview of public transport policy developments in South Africa. *Research in Transportation Economics, 39*, 34–45.

Wan, W. H., Tsang, Y. T., Zhu, H., Koo, C. H., Liu, Y., & Lee, C. C. T. (2018). A real-time drivers' status monitoring scheme with safety analysis. In *IECON 2018—44th Annual Conference of the IEEE Industrial Electronics Society* (pp. 5137–5140). IEEE.

Wilkinson, P. (2006). 'Transit oriented development': A strategic instrument for spatial restructuring and public transport system enhancement in South African cities? In *SATC 2006*.

Yatskiv, I., & Budilovich, E. (2017). A comprehensive analysis of the planned multimodal public transportation HUB. *Transportation Research Procedia, 24*, 50–57.

Zhang, J., Wang, F.-Y., Wang, K., Lin, W.-H., Xu, X., & Chen, C. (2011). Data-driven intelligent transportation systems: A survey. *IEEE Transactions on Intelligent Transportation Systems, 12*, 1624–1639.

Žunić, E., Delalić, S., & Đonko, D. (2020). Adaptive multi-phase approach for solving the realistic vehicle routing problems in logistics with innovative comparison method for evaluation based on real GPS data. *Transportation Letters*, 1–14.

Part II
Policy, Legislative and Governance Frameworks in Urban Public Transport Systems in South Africa and Zimbabwe

Chapter 2
Governance and Structures in Urban Public Transport in South African and Zimbabwean Cities

Abstract Currently, the Fourth Industrial era has accelerated demand for efficiency in urban public transportation governance and is characterized by adaptivity and sustainable societal growth, and economic competitiveness. Furthermore, the global recognition that governance political theory is emerging as dominant in response to disruptive technologies introduced through the Fourth Industrial Revolution, has led government agencies in both the developed and developing world to rebuild the intellectual systems of public transportation. This chapter analyses the international state of what constitutes good policy, and legislative and governance frameworks in urban public transportation systems. The chapter combines a review of policy documents and case studies to better understand the elements of good policy, legislative, and governance frameworks. Consequently, the chapter identifies the lessons learned from literature to aid in the creation of solutions to produce good governance in urban public transportation.

Keywords Governance · Policy · Legislative · Urban public transportation · Fourth industrial revolution

2.1 Introduction

Urban governance is crucial for the functionality of any city and in governing the direction in which development takes place. It is key for all cities to have policies and strategic plans to manage growth and assist urban areas in achieving a high level of competitiveness (Beimborn et al., 2003; Pribyl et al., 2020). However, Public Transport in developing countries such as South Africa is characterized by poor coordination with multiple stakeholders and service providers responsible for providing public transport services for commuters (Sanchez-Atondo et al., 2020; Veeneman & Mulley, 2018). These range from formalized municipal buses in major cities, privately-owned bus companies, rail transport, and non-regulated, informal public transport in the form of mini-bus taxis and tuk-tuks (Mbara, 2016).

Globally many interventions guided by policy and legislation have been proposed to address the challenges of public transportation operations. For example, Emeran et al. (2013) note that public transport is the priority, given the South African context.

T. Gumbo et al., *Urban Public Transport Systems Innovation in the Fourth Industrial Revolution Era*, https://doi.org/10.1007/978-3-030-98717-6_2

Consequently, there should be a transport network of high frequency/volume corridors to shorten travel trips and improve access to economic opportunities. However, Khosa (2017) notes that the country's achievements in urban public transport planning have been disappointing, despite good policy instruments and legislation based on the 1996 white paper, which was aligned with international standards. It is also observable that the apartheid system of governance has contributed to the current land-use pattern which perpetuates long travel distances with location of townships far from major economic nodes (Charman et al., 2017). To bring light to these issues, this chapter examines the international state of what constitutes good policy, legislative, and governance frameworks in Urban Public Transportation systems and presents implications for transport planning before drawing conclusions.

2.2 Conceptual Issues

Urban public transport is critical, as it provides access to markets, employment opportunities, education, health care, and recreation for the economic and social wellbeing of the people. It is also critical in helping the developing world achieve its millennium goals through meeting the needs of poor and vulnerable groups (Hook & Howe, 2005). The state may coordinate such effective operations in the urban public transport system through a regulatory framework, as it has in places like the Rio de Janeiro city region (Santos et al., 2020). Furthermore, reasonable progress has been witnessed in countries like China, where the state has provided adequate policy instruments complemented by the political will to improve public transport (Hu et al., 2010).

Monopoly in space and time by transport operators is another important issue that cannot be ignored, as it may result in some services being offered at higher fares and therefore excluding some poorer members of the public (Anderson et al., 2014). A lack of such comprehensive management by the government of urban public transport operations may result in social and economic exclusion, negatively impact people's quality of life and contribute to market failures (Lucas & Markovich, 2011; Santos et al., 2020). Bahamonde-Birke et al. (2018) also observe that transport planning should be sustainable, where its future impacts are considered when doing trade-offs in desired transport planning outcomes. Politics thus plays a critical role in urban public transport system interventions (Černá et al. 2020). Here, decision-makers shape these systems' direction by enacting innovative policies which redress socio-economic disparities.

According to Sun and Cui (2018), both soft and hard infrastructures must be given cognizance in development processes and projects. Hard infrastructure refers to the railway lines and highways or BRT routes, whereas soft infrastructure implies the secondary institutional or market-oriented development (Nakamura et al., 2017). These observations are also acknowledged in the Gauteng 25-Year integrated transport master plan of 2013 which focuses on intelligent transport systems (Gotz et al., 2014). This outlines the Province's plans to develop sustainable transport systems which will anchor the economic, social, cultural, and environmental objectives of Gauteng.

Investments in urban public transport infrastructure have also emerged as an integral part of improved access to economic opportunities (Assaad et al., 2020). Furthermore, transport infrastructure deficiencies have driven innovative strategies to enhance urban mobility and public transport infrastructure development (Moradi & Vagnoni, 2018). Given the sustainable development agenda and the need for creation of efficient and reliable transport systems, integration of the public transport network has formed the core of the infrastructure for multi-modal transport (Eggleston et al., 2006). This takes account of the socio-economic needs of the people served by the public transport network as well as careful consideration of environmental constraints.

According to Vasconcellos (2014), innovative urban public transport system integration is a response to market failures as well as the need to improve consumer experience. A spatial development pattern of any city is strongly influenced by the transport network which is a result of its planning (Mbatha & Gumbo, 2019). This suggests that a lack of integration in transport planning negatively affects building sustainable, compact, and integrated cities and, therefore, results in market failures. Gotz et al. (2014) observe that cities with vibrant economies have identified innovative urban public transport systems as focus points. Examples of the challenges tackled through innovative public transport systems are traffic congestion problems and increased travel times (Vasconcellos, 2014).

The fragmented nature and spatial form of urban public transport networks operated by various service providers have meant that public institutions' responsibility goes beyond public transport infrastructure. Thus, bringing together all stakeholders who are directly and indirectly involved with public transport infrastructure has improved urban public transport operations. An example of this would be service providers who often use different billing systems, some with distinct technological tools and service patterns, which complicates things for the commuters (Vermesan et al., 2013). Lack of integration and cooperation among service providers leads to increased travel times and costs, as people take more than one mode of transport to get to their destination. This has led to innovative strategies for integrated billing systems.

2.3 Governance Structures and Coordination Mechanisms

The South African public transport system is renowned for its distinct formal and informal public operators (Walters, 2013). The formal public transport system constitutes government-subsidized public transport made up of buses and trains, while the informal public transport network is predominantly the taxi industry, recognized by government but not subsidized (Govender, 2014). Governments have little say over the taxi industry, which is responsible for transporting at least 65% of the public transport commuters in South Africa (Alam & Ahmed, 2013; Govender, 2016).

Furthermore, the taxi industry has played a critical role in the country's public transport arena, especially where the formal public operations like buses and trains could not cater to the people's needs (Lomme, 2008; Sebola, 2014). However, there are many inadequacies, because the taxi industry is often associated with violence; hence the need to create a safer and more convenient public transport system through an innovative urban public transport network has been identified (Govender & Allopi, 2006). The government's previous attempts to formalize and recapitalize or transform the mini-bus taxi industry were later accompanied by the introduction of innovative urban public transport systems (Govender, 2014). The democratic government of South Africa has emphasized socio-economic transformation agenda, widely supported by developed policies and strategies to alleviate poverty, reduce unemployment, and address challenges of inequality, given the ills of the past.

Togo (2016) also observed that a government subsidy has only been offered for the bus system and rail network. This has excluded the taxi industry, arguably the biggest transporter of the urban poor segment of the population which has also contributed to lack of affordability of public transport by the urban poor. Limited access to economic opportunities by the urban poor, where their geographic location is far from workplaces and the cost for transport is increased, means they have to use more than one mode of public transport for a single trip (Gauthier & Weinstock, 2010). This also results in increased traveling times which negatively affect productivity in the workplace and end up discouraging some residents from working, or some even give up their jobs (Rode et al., 2017).

2.4 Interventions on Public Transport Infrastructure

Since 1994, the South African government has embarked on the journey of transforming the country's public transport network and operations through legislative and other initiatives (Togo, 2016). It is evident that the South African Government has prioritized improved public transport through mega investments and strategic policy instruments. Horn (2020) furthermore observes that the South African urban public transport system has reached a critical stage, with major cities (supported by national and provincial governments) already geared up to implement innovative urban public transport systems. Metropolitan cities seem to be driving the innovative urban public transport systems initiatives, while other cities and towns have lagged behind. This is understandable given the population concentration and major economic activities in these metropolitan areas.

The 1996 white paper on national public transport policy is perceived as the key building block that formed the basis of urban public transport systems transformation in the democratic era (Dawood & Mokonyama, 2015). There has been a significant shift in policy instruments since the white paper published in 1996 which further identifies the need for a pro-active approach to deal with the prevailing circumstances and ever-rising challenges facing the country. For instance, the white paper was followed by the Integrated Rapid Public Transport Networks initiative which was a more sophisticated plan for an innovative urban public transport system (Van Ryneveld, 2008). Innovative urban public transport systems like the Gautrain and Bus Rapid Transit (BRT) were also introduced, with the BRT as the commonly adopted form of innovative public transport across the country. The next chapter will discuss this in detail under South African experiences of innovative urban public transport systems.

With the further promulgation of the Infrastructure Act, Act 23 of 2014, South Africa recognized the need to carefully manage and invest in infrastructure development. Consequently, many discussions are taking place around spatial restructuring through integration and building good infrastructure to support the economy. Nevertheless, despite various initiatives and investments in transport infrastructure development, Chirisa et al. (2019) and Chakwizira (2020) argue that poor planning for this growth has led to failures such as severe road traffic congestion, conflicting vehicle–pedestrian movements, increased numbers of uncoordinated small-scale freight vehicles and severe parking shortages. The matrix Table 2.1 depicts the public transport planning, complex issues, policy interventions, and outcomes since 1994.

Despite various initiatives to improve the country's urban public transport system, not much progress has been recorded even after the promulgation of the National Land Transport Act (Dawood & Mokonyama, 2015). This was soon followed by attempts

Table 2.1 Synopsis of the South African urban public transport transformation agenda

Problems associated with transport system, its planning, and operations	Interventions toward transformation of urban public transport operations	Outcomes the initiatives or transformation strategies and prevailing circumstances
• Spatial fragmentation due to past land-use planning patterns • Rapid population growth and densely populated previously disadvantage communities • High costs and long travel hours affect mostly the urban poor • Lack of cooperation from transport operators • Poor ticketing system where commuters had to pay for every mode of transport they use • Increased number of private transport usage • Informal and non-regulated public transport operations • Poor subsidy targeting	• 1995 White Paper on National Transport Policy which led to a Green Paper in March 1996 • The National Land Transport Transition Act (Act 22 of 2000) • The National Land Transport Transition Amendment Act (no 26 of 2006) (NLTTAA) • Integrated Rapid Public Transport Networks • Accelerated Modal Upgrading • Integrated Rapid Public Transport Networks (IRPTNs) • Gautrain and Bus Rapid Transit • Subsidy Scheme to target low-income earners	• Introduction of innovative urban public transport systems in form of Bus Rapid Transit in major cities • Gautrain connecting mainly the three Gauteng Metropolitan Cities major transport nodes • Increased public transport subsidy expenditure without balanced benefits by the previously disadvantaged and the general public • Mini-buses or taxis still not covered by the public subsidy • BRT and Gautrain having own subsidy requirements • Innovative urban public transport systems not easily accessible and not affordable to the previously disadvantaged communities

to recapitalize minibus taxis and restructuring of bus operations contracts under the accelerated modal upgrade program (Govender, 2014; Sebola, 2014). Nevertheless, Dawood and Mokonyama (2015) observe that the South African public transport challenges have been persistent despite various transformation and formalization initiatives since 1996. The need for stakeholder cooperation has been identified as a crucial component that is lacking, despite continued massive investments for public transport infrastructure and services by government (Department of Transport, 2017; May et al., 2017; Walters, 2013). Thus, a need exists to strengthen the public sector human resource with carefully analyzed and understood public transport markets trends in order to influence the country toward a comprehensive urban public transport network and implementation of policy and legislation.

2.5 Gauteng 25-Year Integrated Transport Master Plan (2013)

The South African government at all levels is trying out innovative ideas to improve public transport systems. The introduction of the Gautrain system and BRT system in the Gauteng province was a good initiative, as this has met other challenges of urban public transport in the province. The province's policies and legislative frameworks also include more efficient integrated systems and consider the use of technology in these operations. Attesting to the above, the 25-year Integrated Transport Master Plan (ITMP25) contains a full implementation strategy for the transformation of the transport system in Gauteng over the next 25 years. However, given the current state of transport in the Gauteng City Region and the associated pressing problems and challenges, some urgent interventions are required. A few initiatives have been identified to integrate urban public transport in the Gauteng province. In regard to the study, integrated public transport ticketing and information dissemination are identified as one of the key initiatives.

2.6 Unpacking the Role of Governance and Structures in Urban Public Transport

The primary objective of public transport infrastructure is to facilitate urban mobility by way of improving the movement of people, goods, and services within a particular space (Department of Transport, 2017). Given the prominent spatial disparities in the developing world, innovative transport systems policy interventions must seek to improve urban mobility by effectively facilitating spatial integration of various land uses, thereby creating a vibrant socio-economic environment (Jennings, 2015). Thus, densification along main public transport corridors has been advocated as a mechanism to redress the spatial imbalances of the past (Wilkinson, 2006). Even in the developed world, rapid urbanization and population growth have meant that adequate public transport is a top priority for urban growth management.

In a nutshell, urban mobility is concerned with the time spent traveling, and the distance covered is central to the state of mobility within the urban environment (Gotz et al., 2014). However, Musakwa and Gumbo (2017) argue that urban mobility policy interventions should look beyond addressing traffic problems. They should be responsive to people's day-to-day needs and not neglect or systematically exclude the urban poor. Accordingly, when deliberating on innovative urban transport systems, one cannot neglect the demand for improved urban mobility characterized by improved access to opportunities, affordable public transport systems, and efficient and reliable service at the core of improving urban public transport infrastructure (Ndwandwe & Gumbo, 2020).

Thus, the existing body of knowledge reveals four interlinked and interdependent main themes that serve as performance indicators of the public transport systems. These are accessibility, affordability, efficiency, and reliability or convenience (Moyo et al., 2018). Urban public transport systems that reflect a balance across these four main themes can be deemed to have met the minimum requirements for improved urban mobility through an integrated urban public transport network (Togo, 2016). Subsequently, innovative urban public transport systems policy interventions should strive to achieve these main principles to ensure fully functional and economically viable cities.

2.6.1 Accessibility Through Urban Public Transport Systems

The day-to-day activities within an urban setting consist of varied social and economic traveling needs, where people require easy access to economic opportunities and other urban amenities (Culwick et al., 2015). Such accessibility is concerned with people's ability to easily connect to various opportunities in the form of goods and services in a given city, having reduced travel times and walkable access to human settlements as key indicators of improved urban mobility (Gwala, 2007). However, central to attaining accessibility is the provision of a variety of modes of public transport that are operationally integrated to enhance connectivity of various nodes. Thus, Commuters can travel directly to their desired destination without experiencing unnecessary delays due to limited public transport options.

Nevertheless, improved accessibility through public transport cannot be achieved if public transport systems neglect people's needs and travel costs (Gotz et al. 2014). In relation to people's needs, therefore, the principle of accessibility goes beyond access to economic opportunities. It emphasizes the importance of providing public transport systems that are user-friendly to the elderly and people living with disabilities (Chowdhury et al., 2018). Non-Motorized Transport, through pedestrian corridors and cycling routes, also forms the core of accessibility through innovative urban public transport systems (Lesh, 2013). Such systems should facilitate all travel needs through public transport corridors that bring socio-economic vibrancy to city life.

In spite of these shortfalls observed in traditional approaches to public transport in relation to the mobility principle of accessibility, improvements in the efficacy of innovative urban public transport systems will improve access levels. Innovative approaches have been adopted which are essential for attainment of accessibility. Through these, all people regardless of their physical, economic, and social backgrounds are able to move around and enjoy opportunities available within the urban setting. Innovative approaches examine all components that are central to achieving a fully functional urban public transport network. This makes it easier for people, goods, and services to move, while also fostering integration of economic and residential opportunities.

2.6.2 *Affordability: Cost-Effective Urban Public Transport Systems*

In the developing countries characterized by a fragmented spatial form, modes of public transport do not take commuters directly to their respective place of work or residence. As a result, this necessitates commuters' use of two or more modes of public transport before getting to their intended destination (Yatskiv & Budilovich, 2017). Further, the urban poor is often the main users of public transport, but due to their financial constraints, they are unable to afford public transport fares and are forced to walk long distances to obtain access to various opportunities (Beukes et al., 2011). Such increased travel times and costs have a negative impact on the productivity of workers, and in some cases, these circumstances may lead to low-income earners (the urban poor) giving up their jobs or being discouraged from looking for employment opportunities completely.

Urban mobility policy interventions through innovative urban public transport systems must therefore see to it that affordability for all is enhanced, especially for the poor (May et al., 2017). Since the colonially inherited spatial form in most developing countries cannot be reversed, integrated billing systems and cooperation among stakeholders may help improve the situation. For example, in Rio de Janeiro (Brazil), introduction of the integrated billing system of one travel card Bilhete Único Integrado (BUI) or Integrated Ticket Fare meant that commuters could use one card to pay for all modes of public transport (Mbatha & Gumbo, 2019; Santos et al., 2020). As such, this helped facilitate affordable multi-nodal public transport operations and ensured reduced travel times which were very beneficial to low-income earners.

2.6.3 *Reliability of Public Transport Systems*

Urbanization and rapid population growth mean that availability of reliable public transport systems is of paramount importance to avoid congestion on public roads

(Guzman et al., 2020). In the twenty-first century, people are not easily persuaded to get rid of private cars unless they can fully rely on public transport for their social and economic needs without having to endure inconvenience (Rode et al., 2017). Thus, people's having to rely on urban public transport for commuting purposes while being concerned about the punctuality of such systems cannot boost commuters' confidence in public transport (Togo, 2016).

In response to this, Intelligent Transport Systems (ITS) have been introduced to strengthen the reliability of urban public transport systems through constant updates and communication about such services (Mangiaracina et al., 2017). Through ITS communication channels, people are able to plan their business and day-to-day trips through the information provided by the channels (Zhang et al., 2011). ITS even enables tourists to navigate through the city using public transport. They can go wherever they want while being guided by the ITS linked to innovative urban public transport systems. These clearly communicate the mode of public transport to take them conveniently to their destination.

2.7 Implications, Reflections, and Generalizability of Findings

There are several implications drawn from this chapter with regard to the new knowledge generation for academic purposes, and policy formulation and implementation in the local, provincial, and national spheres. In particular, the implementation of various transport integration solutions may result in reduction of travel times, transportation costs, environmental pollution, and traffic congestion. Transport integration leads to the improvement of urban public transport system accessibility and overall competitiveness. It assures better utilization of different transportation means and infrastructure. Moreover, integration of urban public transportation is mostly determined by the pattern of land-use, the nature of the transportation systems, and the characteristics of the traveler. Travel cost, time, distance, and the choice of travel mode are all important. Most importantly, the closer the origin and destination are to the main transportation system the higher the level of connectivity. Also, the wider the variety of modes for traveling between a given origin and a particular destination, the greater the connectivity. In addition, less time and money spent on travel means that more places can be reached within a certain budget, and the greater the connectivity. However, in order for the concept of connectivity to be useful for evaluation of the need for and effectiveness of transportation and land-use planning policies, it needs to be translated into measures of connectivity.

To ensure improved urban mobility, innovative urban public transport systems should be anchored through application of analytical methods to achieve best practices. These should be characterized by best consumer experience and socio-economic vibrancy, thereby optimizing efficiency of the public transport systems. Therefore, efficiency of public transport systems is strongest where service providers see to it that service is of a high-quality. Moreover, the safety and security of commuters should be a priority, and each mode of public transport should be consumer-friendly and effectively managed. However, modes of public transport cannot run efficiently when operating independently from each other. Consequently, an integrated multi-modal urban public transport network is the heartbeat of efficiency, as it enables people to connect to every part of the city through high-quality public transport systems.

Spatial integration is central to the principle of efficiency, where various innovations can be undertaken to facilitate trade, tourism, commerce, services, and education to benefit commuters while also boosting the local economy. Therefore, efficiency goes beyond the quality of service provided by urban public transport systems. It also extends to designing stations and connector points as activity nodes, either as green space plazas for leisure and recreational purposes, or socio-economic nodes with vibrant small-scale economic activities (Han & Fang, 2000). At an advanced level, efficiency can even involve a combination of recreational and socio-economic activities, where people can relax, enjoy and operate their small businesses in a healthy, refreshing, and vibrant environment.

Where urban areas are characterized by a fragmented spatial form, as they are in South Africa, access to a fully functional mode of public transport that provides reliable and affordable service to all income groups contributes to improved accessibility of opportunities. Nevertheless, findings of this study reveal that urban communities in South Africa face substantial and real challenges when it comes to transportation and access to economic opportunities (Culwick et al., 2015). Hence innovative urban transport systems have become the focal point of discussion to improve the situation. With major investments in rail and road transport in South Africa, the expectation is that such investments should translate to spatial integration and economic transformation.

Consequently, the essential role of innovative urban public transport systems has already been established throughout this discussion chapter. In this regard, transport planning has been identified as a key sector in facilitating improved urban mobility and socio-economic transformation and transformation. While this has been said, it is also important to note that places are unique, and area-specific issues should be studied, analyzed, and understood in order to devise relevant solutions or recommendations. South Africa's prevailing socio-economic disparities and uniqueness of different areas within the country should be noted and attended to. However, trends from innovation in public transportation suggest that massive investments in public transport infrastructure have placed little emphasis on the creation of economic opportunities for the urban poor through encouragement of small-scale entrepreneurs.

The chapter reveals that a lack of collaboration among stakeholders (including state entities) where each institution works independently toward its core mandate without being in full cooperation with other interrelated and interdependent decision or competency areas has meant a loss of opportunity. This was evident through limited or no involvement of the department of local economic development in implementing innovative urban public transport systems. These could have possibly included Small,

Medium and Micro-sized Enterprises (SMMEs) business needs as part of the implementation of innovative urban public transport systems. For instance, the Hatfield Station was the obvious example of a missed opportunity for creating activity nodes for social and economic interactions that encourage small-scale entrepreneurship and informal economy.

2.8 Conclusion

It is evident that transport planning has evolved from a more traditional toward a rational, comprehensive, and integrated approach to improve urban mobility and enhance economic growth. Through governance and legislative structures, an integrated approach to innovative urban public transport systems and socio-economic transformation will ensure cities can become foci for innovation, where trade, tourism, business operations, services delivery, access to opportunities, and education can be enhanced. Furthermore, an enabling environment for small-scale entrepreneurship growth and employment opportunities should create a trickle-down effect when implementing urban public transport systems. Such opportunities must see to the transformation of the polarized labor market. To achieve this, human rehabilitation and skills development are essential among the previously disadvantaged.

Given the findings of this study, socio-economic transformation agenda still need to deliberately facilitate the inclusion of the urban poor in line with the massive public transport infrastructure investments. Results have shown that the urban poor and previously disadvantaged are yet to benefit from innovative urban public transport systems investments meaningfully. Major investments should prioritize the urban poor instead of only improving the lives of those who are already part of the formal economy. Of paramount importance is fostering investment of returns and positive spinoffs through socio-economic benefits which lead to economic growth and development. To achieve these, an innovative urban public transport system must be coordinated through a collaborative planning process, and an analysis of interconnected decision areas to achieve improved mobility and socio-economic transformation.

Mechanisms must therefore be developed to ensure that transport planning investments and decisions do not compromise the urgent need for socio-economic transformation, i.e., focus must be on realistic solutions to community problems. For example, there are more private and municipal buses required which can transport workers and students during certain hours in the morning and late afternoon to the evening. Investment in the development of interventions led by policy and legislative instruments will improve the quality of life of commuters through facilitating affordable and safe public transport services to reduce the daily travel times between home and work. Furthermore, non-integrated transport planning across various modes has resulted in modes that are not sufficiently customer-focused, that are inefficient, and have poor levels of reliability, predictability, comfort, and safety. Such planning does not reflect the world-class aspirations of policy and legislative instruments. The fragmented nature of institutional governance over public transport is also not helpful. Consequently, policies and legislative frameworks must support the integration of urban public transport, as their strategies of improving the public transport focus on integration.

References

Alam, M. A., & Ahmed, F. (2013). Urban transport systems and congestion: A case study of Indian cities. *Transport and Communications Bulletin for Asia and the Pacific, 82*, 33–43.

Anderson, M., Wilson, J., & Khan, T. (2014). Evaluation of origin-destination matrix estimation techniques to support aspects of traffic modeling.

Assaad, R., Dagli, C., & El-Adaway, I. H. (2020). A system-of-systems model to simulate the complex emergent behavior of vehicle traffic on an urban transportation infrastructure network. *Procedia Computer Science, 168*, 139–146.

Bahamonde-Birke, F. J., Kickhöfer, B., Heinrichs, D., & Kuhnimhof, T. (2018). A systemic view on autonomous vehicles: Policy aspects for a sustainable transportation planning. *disP-The Planning Review, 54*, 12–25.

Beimborn, E. A., Greenwald, M. J., & Jin, X. (2003). Accessibility, connectivity, and captivity: Impacts on transit choice. *Transportation Research Record*.

Beukes, E., Vanderschuren, M., & Zuidgeest, M. (2011). Context sensitive multimodal road planning: A case study in Cape Town, South Africa. *Journal of Transport Geography, 19*, 452–460.

Černá, L., Zitrický, V., & Abramović, B. (2020). Methodical manual for a set of transport regulations in railway passenger transport. *LOGI–Scientific Journal of Transport and Logistics, 11*, 13–24.

Chakwizira, J. (2020). The intersection of spatial fragmentation and smart transport planning in Gauteng Province, South Africa: Constraints and opportunities. In *International Conference on Applied Human Factors and Ergonomics* (pp. 367–373). Springer.

Charman, A., Tonkin, C., Denoon-Stevens, S., & Demeestere, R. (2017). *Post-apartheid spatial inequality: Obstacles of land use management on township micro-enterprise formalisation.* A report by the Sustainable Livelihoods Foundation. Sustainable Livelihoods Foundation.

Chirisa, I., Mhlanga, G., & Matamanda, A. R. (2019). Infrastructure, city-region development, and Africa's territorial spaces: Gauteng as the exemplar. In *Optimizing regional development through transformative urbanization.* IGI Global.

Chowdhury, S., Hadas, Y., Gonzalez, V. A., & Schot, B. (2018). Public transport users' and policy makers' perceptions of integrated public transport systems. *Transport Policy, 61*, 75–83.

Culwick, C., Gotz, G., Katumba, S., Trangoš, G., & Wray, C. (2015). Mobility patterns in the Gauteng City-Region, South Africa. *Regional Studies, Regional Science, 2*, 309–311.

Dawood, G., & Mokonyama, M. (2015). Towards a more optimal passenger transport system for South Africa: Design of public transport operating subsidies. *Council for Scientific and Industrial Research (CSIR), 62*, 192–213.

Department of Transport. (2017). Draft Revised White Paper on National Transport Policy.

Eggleston, H., Buendia, L., Miwa, K., Ngara, T., & Tanabe, K. (2006). 2006 IPCC guidelines for national greenhouse gas inventories.

Emeran, H., Heyns, W., & Dyer, R. (2013). PRASA's new National Strategic Plan: Transportation engineering. *Civil Engineering= Siviele Ingenieurswese, 2013*, 12–15.

Gauthier, A., & Weinstock, A. (2010). Africa transforming paratransit into BRT. *Built Environment, 36*, 317–327.

Gotz, G., Wray, C., Venter, C., Badenhorst, W., Trangoš, G., & Culwick, C. (2014). Mobility in the Gauteng city-region.

Govender, K. K. (2014). Public transport service quality in South Africa: A case study of bus and mini bus services in Johannesburg. *African Journal of Business Management, 8*, 317–326.

Govender, K. K. (2016). Exploring public transport service quality: The case of mini-bus taxi service in South Africa. *Eurasian Business Review, 6*, 101–116.

Govender, R., & Allopi, D. (2006). Towards a safer minibus taxi industry in South Africa. In *SATC 2006.*

Guzman, L. A., Arellana, J., & Alvarez, V. (2020). Confronting congestion in urban areas: Developing Sustainable Mobility Plans for public and private organizations in Bogotá. *Transportation Research Part a: Policy and Practice, 134*, 321–335.

Gwala, S. (2007). Urban non-motorised transport (NMT): A critical look at the development of Urban NMT Policy and planning mechanisms in South Africa from 1996–2006. In *SATC 2007.*

Han, X., & Fang, B. (2000). Four measures of transportation's economic importance.

Hook, W., & Howe, J. (2005). *A background paper to the task force on slum dwellers of the millennium project.* ITDP.

Horn, A. (2020). The darker side of muddling through: An analysis of spatial planning policy decision-making and policy termination in Cape Town, South Africa. In *Urban Forum* (pp. 1–26). Springer.

Hu, X., Chang, S., Li, J., & Qin, Y. (2010). Energy for sustainable road transportation in China: Challenges, initiatives and policy implications. *Energy, 35*, 4289–4301.

Jennings, G. (2015). A bicycling renaissance in South Africa? Policies, programmes & trends in Cape Town. In *Southern African Transport Conference.*

Khosa, M. (2017). Public transport in the changing South Africa, 1994–2000. In *Geography and economy in South Africa and its neighbours* (pp. 146–168). Routledge.

Lesh, M. C. (2013). Innovative concepts in first-last mile connections to public transportation. In *Urban public transportation systems 2013*

Lomme, R. (2008). Should South African minibus taxis be scrapped? Formalizing informal urban transport in a developing country. In *Proceedings of the CODATU XIII Conference*, Ho Chi Minh City.

Lucas, K., & Markovich, J. (2011). International perspectives. In *New perspectives and methods in transport and social exclusion research.*

Mangiaracina, R., Perego, A., Salvadori, G., & Tumino, A. (2017). A comprehensive view of intelligent transport systems for urban smart mobility. *International Journal of Logistics Research and Applications, 20*, 39–52.

May, A., Boehler-Baedeker, S., Delgado, L., Durlin, T., Enache, M., & van der Pas, J.-W. (2017). Appropriate national policy frameworks for sustainable urban mobility plans. *European Transport Research Review, 9*, 7.

Mbara, T. C. (2016). Tuk-tuk, 'new kid on the block' in Johannesburg: Operational and user travel characteristics, competition and impacts. *Journal of Transport and Supply Chain Management, 10*(1), 1–9.

Mbatha, S., & Gumbo, T. (2019). Establishing the state of spatial integratedness of innovative public transport systems in Gauteng, South Africa. In *REAL CORP 2019–IS THIS THE REAL WORLD? Perfect smart cities vs. real emotional cities. Proceedings of 24th International Conference on Urban Planning, Regional Development and Information Society* (pp. 319–327). CORP–Competence Center of Urban and Regional Planning.

Moradi, A., & Vagnoni, E. (2018). A multi-level perspective analysis of urban mobility system dynamics: What are the future transition pathways? *Technological Forecasting and Social Change, 126*, 231–243.

Moyo, T., Musakwa, W., & Mokoena, B. (2018). An analysis to investigate spatial cognitive factors which influence cycling patterns in Johannesburg. *International Archives of the Photogrammetry, Remote Sensing & Spatial Information Sciences.*

Musakwa, W., & Gumbo, T. (2017). *Impact of urban policy on public transportation in Gauteng, South Africa: Smart or dumb city systems is the question.* Springer.

Nakamura, F., Makimura, K., & Toyama, Y. (2017). Perspective on an urban transportation strategy with BRT for developing cities. *Engineering and Applied Science Research, 44*, 196–201.

Ndwandwe, B., & Gumbo, T. (2020). Vibrant or dull urban spaces: Are city of Tshwane "A Re Yeng" and "Gautrain" connector points places of social and economic interaction? In *REAL CORP 2020—SHAPING URBAN CHANGE LIVABLE CITY REGIONS FOR THE 21st CENTURY. Proceedings of 25th International Conference on Urban Planning and Regional Development in the Information Society*.

Pribyl, O., Blokpoel, R., & Matowicki, M. (2020). Addressing EU climate targets: Reducing CO_2 emissions using cooperative and automated vehicles. *Transportation Research Part D: Transport and Environment, 86*, 102437.

Rode, P., Floater, G., Thomopoulos, N., Docherty, J., Schwinger, P., Mahendra, A., & Fang, W. (2017). Accessibility in cities: transport and urban form. *Disrupting Mobility*, 239–273.

Sanchez-Atondo, A., Garcia, L., Calderon-Ramirez, J., Gutiérrez-Moreno, J. M., & Mungaray-Moctezuma, A. (2020). Understanding public transport ridership in developing countries to promote sustainable urban mobility: A case study of Mexicali, Mexico. *Sustainability, 12*, 3266.

Santos, T., Silva, M. A., Fernandes, V. A., & Marsden, G. (2020). Resilience and vulnerability of public transportation fare systems: The case of the city of Rio De Janeiro, Brazil. *Sustainability, 12*, 647.

Sebola, M. (2014). Recapitalizing mini-bus taxis for effective public transportation in South Africa: The urban rural transport connection problem. *Urban Transport XX, 138*, 125.

Sun, Y., & Cui, Y. (2018). Evaluating the coordinated development of economic, social and environmental benefits of urban public transportation infrastructure: Case study of four Chinese autonomous municipalities. *Transport Policy, 66*, 116–126.

Togo, M. (2016). Adoption of green technologies in the public transport sector: The case of Rea Vaya BRT System, South Africa. *Milestones in Green Transition and Climate Compatible Development in Eastern and Southern Africa*, 93.

Van Ryneveld, P. (2008). *15 year review of public transport in South Africa with emphasis on metropolitan areas*. Unpublished paper commissioned for the Fifteen Year Review.

Vasconcellos, E. A. (2014). *Urban transport environment and equity: The case for developing countries*. Routledge.

Veeneman, W., & Mulley, C. (2018). Multi-level governance in public transport: Governmental layering and its influence on public transport service solutions. *Research in Transportation Economics, 69*, 430–437.

Vermesan, O., Friess, P., Guillemin, P., Sundmaeker, H., Eisenhauer, M., Moessner, K., Le Gall, F., & Cousin, P. (2013). Internet of things strategic research and innovation agenda. In *Internet of things: Converging technologies for smart environments and integrated ecosystems*. River Publishers.

Walters, J. (2013). Overview of public transport policy developments in South Africa. *Research in Transportation Economics, 39*, 34–45.

Wilkinson, P. (2006). 'Transit oriented development': A strategic instrument for spatial restructuring and public transport system enhancement in South African cities? In *SATC 2006*.

Yatskiv, I., & Budilovich, E. (2017). A comprehensive analysis of the planned multimodal public transportation HUB. *Transportation Research Procedia, 24*, 50–57.

Zhang, J., Wang, F.-Y., Wang, K., Lin, W.-H., Xu, X., & Chen, C. (2011). Data-driven intelligent transportation systems: A survey. *IEEE Transactions on Intelligent Transportation Systems, 12*, 1624–1639.

Chapter 3
Role of Policy and Legislative Instruments in South African and Zimbabwean Cities

Abstract The Fourth Industrial Revolution commonly denoted as 4IR has brought with it an increased demand for innovative public transportation systems; notable examples include carpooling services, rapid railway systems, rapid bus systems, and green mobilities. Traditionally, the role of policy and legislative frameworks was to be the cornerstone of facilitating transformation, integration, change, and reform to aide functionality and economic prosperity of cities and towns. Academically, literature on the role of policy and legislative frameworks to govern emerging public transportation systems during the Fourth Industrial Revolution is increasing. However, there is still a knowledge gap, regarding the readiness of policy and legislative frameworks in developing countries to facilitate the growth of innovative public transportation. This chapter builds on lessons learned from the previous chapter by seeking to explore and deliberate on the role of current policy and legislative instruments to facilitate the growth of innovative public transportation systems in the era of the Fourth Industrial Revolution in cities in the developing world. To unpack this role, we use a systematic method combining bibliometric and literature review analysis. Preliminary findings reveal a lack of a long-term implementation horizon coupled with a lack of stakeholder cooperation and integration within and across government spheres. Policy and legislative frameworks, although they are in support of innovative public transportation, fail to be all-inclusive, as they only play a supporting role to innovation, with most innovative initiatives originating from the private sector.

Keywords Policy and legislative frameworks · Public transport systems · Fourth industrial revolution · Southern Africa

3.1 Introduction

Public Sector driven capital investment initiatives are often based on clearly defined policy directives and laws. For such public transport infrastructure investments, economic growth and development have been the focus of all policy instruments and legislative frameworks. They have also been regarded as pivots of functionality of economic nodes and communities. Thus, innovative strategies for investment in public transport infrastructure have gained momentum as contributors of spatial

T. Gumbo et al., *Urban Public Transport Systems Innovation in the Fourth Industrial Revolution Era*, https://doi.org/10.1007/978-3-030-98717-6_3

integration and socio-economic transformation through improving urban mobility, while simultaneously facilitating business and small-scale entrepreneurship growth and employment creation. The commonly envisaged outcome of such innovative approaches is a spatially integrated urban form through a multi-modal public transport network and creation of vibrant urban centers and economic spaces characterized by a variety of social and economic activities.

In metropolitan cities throughout South Africa and Gauteng, it is evident that commendable work has been done through investments in public transport infrastructure (Makhura, 2015). However, approaches adopted for planning and implementation of projects in line with policy directives on public transport infrastructure leave much to be desired, despite massive investments. Furthermore, spatial connectivity and socio-economic transformation still need to be deliberated upon to facilitate inclusion of the urban poor in line with the considerable public transport infrastructure investments. The urban poor and previously disadvantaged communities are yet to benefit meaningfully from these innovative urban public transport systems investments. Thus, major investments should prioritize the urban poor instead of mainly improving the lives of those who are already part of the formal economy.

Public transport planning has evolved from a more traditional approach toward a rational, comprehensive, and integrated discipline to improve urban mobility and socio-economic transformation. Through this, contemporary cities can become hubs for innovation through an integrated approach to innovative urban public transport systems. Currently, cities in Southern Africa's approach to development planning still apply a top-down approach (Dubina et al., 2017). As such, numerous governments develop frameworks through policy directives that provincial and municipal departments must work toward. These provincial departments are then tasked with the responsibility of regulating provincial-specific matters. The result of this system is that, although policy directives are set at national level, with provinces setting legislative frameworks and policy directives in line with provincial differentiation, the actual implementation takes place within the Municipal sphere. The general expectation is therefore that the three spheres must work together, with the national and provincial spheres playing a supportive and monitoring role. There is also no exception with the public transport infrastructure, where the national government has been at the center of setting the tone and direction for growth and development.

For this chapter, it was important for the researchers to investigate the effectiveness of the commonly followed top-down approach to policy and legislative interventions. Thus, the first point of departure was looking at the characteristics of the Southern African urban public transport systems and infrastructure adequacy. This chapter, therefore, gives an overview of the Southern African urban public transport systems with a focus on spatial and socio-economic imbalances. Southern African public transport operations and government interventions in public transport infrastructure to redress the said past imbalances are then discussed based on government resource documents and the existing body of knowledge.

Subsequently, the chapter outlines the effectiveness of government legislative frameworks and policy interventions. A case study of the Gauteng province is also used to outline specific policy interventions in public transport infrastructure while also reviewing the level of cooperation between the provincial and national departments and between provincial and municipal departments. This is essential since the provincial government is the link between the national government that sets out policy directives and the municipal sphere where actual implementation occurs. The chapter concludes with a synopsis of the Gauteng Province innovative urban public transport systems in conjunction with national and provincial legislative frameworks and policy directives.

3.2 Innovative Urban Public Transport Systems

Urban public transport systems are arguably the lifeblood of cities and towns, as centers of interaction and economic activities essential for connecting places and people (Mazaza, 2002). The fact that urban public transport systems are the heartbeat of urban areas means that, for urban development planning to meet its intended need effectively, provision of adequate public transport infrastructure should be a top priority for public institutions and cities worldwide. Thus, the pressing need exists for having an efficient urban public transport system that addresses socio-economic needs of the people through an integrated urban public transport network. This has influenced the nature of urban planning for developing countries which have the challenge to manage rapid urbanization against their constrained and at times decaying public transport infrastructure (Albalate & Bel, 2009).

Notably, urban public transport systems play a critical role in transporting high volumes of people for varied day-to-day transportation needs, access to economic prospects, and transporting goods (Sodhro et al., 2019). Through a fully functional and efficient urban public transport system, cities can promote access to economic opportunities and minimize congestion and pollution, where public transport ineffectiveness hampers citizens' quality of life and productivity (eThekwini Municipality, 2016). Moreover, urban public transport systems have earned a reputation for solving congestion and road traffic management. Therefore, fully functional, efficient urban public transport systems reduce congestion on public roads, where an alternative is provided and where people reach their respective destinations timeously.

People have different travel patterns and preferences which influence public transport network patterns and make possible the transfer between different modes of transport. The common multi-modal systems of a public transport network include bus services (including mini-buses or taxis), rail transport, subways, and cycling, with the main alternatives being bus systems and rail transport. Commonly, there are different service providers in major urban areas, some of which can be formal or informal, however, buses and rail transport are the most common modes of urban public transport systems (Vuchic, 2002). Consequently, it is not surprising that Bus Transit Systems (BRT) and advanced train systems have formed the core of innovative public transport over the past 20 years (Wright, 2014).

Urban public transport is key, as it provides access to markets, employment opportunities, education, health care, and recreation for the wellbeing of the people economically and socially (Ministry of Transport, 2019). Improved urban public transport is also critical in helping the developing world achieve its millennium goals, as it also caters to poor and vulnerable groups (Hook & Howe, 2005). Thus, through a regulatory framework, the state may unify and coordinate operations in the urban public transport system, as it has in places such as Rio de Janeiro city (Zottis, 2014). Progress has also taken place in countries like China, where the state has provided adequate policy instruments complemented by the political will to improve public transport systems (Jones et al., 2014).

The monopoly in space and time by transport operators is an issue that cannot be ignored, as it may result in some services being offered at higher fares and therefore excluding some members of the public, especially the urban poor (Anderson & Wilson, 2014). This means that a lack of comprehensive management and interventions by the government in urban public transport operations may result in social and economic exclusion and negatively impact people's quality of life (Markovich & Lucas, 2011). A lack of regulatory measures and enforcement, therefore, contributes to market failures (Den Hertog, 2010). Thus, transport planning should be done in a sustainable manner, where its future impacts are considered when doing trade-offs in transport planning desired outcomes (Litman, 2012). Cerna (2013) also argues that political will plays a critical role in urban public transport systems interventions, where decision-makers play a critical role in shaping the direction of urban public transport systems by enacting innovative policies which redress socio-economic disparities.

According to Mulenga (2013), both soft and hard infrastructures must also be taken cognizance of in development processes and projects. Hard infrastructure relates to the railway lines and highways or BRT routes, whereas soft infrastructure refers to institutional or market-oriented development, secondary to hard infrastructure (Fung et al., 2006). These observations are also acknowledged in the Gauteng 25-Year integrated transport master plan of 2013 which focuses on intelligent transport systems. This plan outlines the province's envisaged plans to develop sustainable transport systems that will anchor the economic, social or cultural, and environmental objectives of Gauteng.

Investments in urban public transport infrastructure have thus emerged as an integral part of socio-economic transformation and improved access to economic opportunities (Marrian, 2001). Existing deficiencies in urban public transport infrastructure have driven innovative strategies or approaches to urban mobility and public transport infrastructure development (Gakenheimer, 1999). Given this sustainable development agenda and the need for efficient and reliable urban public transport systems, integration of the public transport network has formed the core of urban public transport infrastructure for multi-modal transport (Page, 2012). These consider carefully the socio-economic needs of the people served by the public transport network as well as environmental constraints.

According to the Danish Business Authority (2012), innovative urban public transport system integration responds to market failures as well as improved consumer experience. A spatial development pattern of any city is strongly influenced by the transport network which is a result of its planning (Clark, 2000). Therefore, a lack of integration in transport planning negatively affects building of sustainable, compact, and integrated cities and results in market failures. Pardo et al. (2010) observed that cities with higher ranking and vibrant economies have identified innovative urban public transport systems as focus points. Traffic congestion problems resulting in increased travel times are therefore challenges that can only be tackled through innovative public transport systems (Vasconcellos, 2004).

The urban public transport network operated by various service providers together with the spatial form in the development is currently fragmented. This has meant that public institutions' responsibility goes beyond public transport infrastructure. This need to bring together all stakeholders, directly and indirectly, involved with public transport infrastructure has led to improvement in urban public transport operations. The different service providers often use different billing systems service patterns, some with distinct technological tools and service patterns, which complicates things for commuters (Vermesan & Friess, 2013). Such lack of integration and cooperation among service providers leads to increased travel times and costs, as people take more than one mode of transport to get to their destination, and this has led to innovative strategies for integrated billing systems (Secure Technology Alliance, 2017).

3.3 Typology of Innovative Urban Public Transport Systems

The need has been established for innovative approaches to planning and implementing urban public transport systems and provision of adequate public transport infrastructure that will enable improved quality of public transport services (Ndwandwe & Gumbo, 2017). Because of this, it is important to understand the characteristics of innovative urban public transport systems in their varied forms and then derive adequate policy instruments such that investments are channeled accordingly while considering the public transport needs of that specific city (Karim, 2017). Hence, innovative public transport systems can be defined as new practices that provide better solutions to existing challenges and traditional measures in urban transport planning and mobility (Polis, 2015).

Innovative urban public transport systems allow for interdependent stakeholders and public transport service providers to collectively respond to public transport challenges (Page, 2012). In this regard, the end goal should be a multi-modal public transport network that brings together the operations of all public transport modes that are responsive to local issues and conditions (Karim, 2017; Litman, 2014). Such innovations range from technologically inclined, high-quality passenger transportation, stakeholders' cooperation, and integrated billing systems across all modes of public transport systems. The common goal is always providing, efficient, reliable, and affordable urban public transport systems.

The said innovative urban public transport systems characterized by rapid rail transport, light rail transport, and Bus Rapid Transit (BRT) systems have since been the major investment pillars for most governments at local and international levels (Tsay & Herrmann, 2013). The major investments associated with these innovations are understood to be key drivers of economic growth and social development. Therefore, it is important to understand each type of innovative urban public transport system, the type and quantity of passengers it carries, and the role they play in socio-economic transformation, given prevailing circumstances in a city or country. The most important consideration for developing countries is ensuring that they do not opt for public transport options that are too costly for them to implement; rather they must seek innovations that are within their financial capabilities.

3.4 Innovative Public Transport Systems and Urban Mobility

The primary objective of public transport infrastructure is to facilitate urban mobility by way of improving the movement of people, goods, and services within a particular space (Department of Transport, 1996). Given the prominent spatial disparities in the developing world, it is important that innovative urban public transport systems policy interventions seek to improve urban mobility. This is achieved by effectively facilitating spatial integration of various land uses thereby creating a vibrant socio-economic environment within cities or towns (Jennings, 2015). Thus, densification along main public transport corridors has been advocated as a mitigation mechanism to redress the spatial imbalances of the past (Wilkinson, 2006). Even in the developed world, rapid urbanization and population growth have meant that adequate public transport has been made a top priority for urban growth management and development facilitation (Aljoufie et al., 2011).

Urban mobility is concerned with the time spent traveling, and distance covered is central to the state of mobility within the city (Lucas et al., 2009). However, Peng (2005) argues that urban mobility policy interventions and investments should look beyond addressing issues of congestion and traffic problems; they should be responsive to people's day-to-day needs and not neglect or systematically exclude the urban poor. Accordingly, when deliberating on innovative urban transport systems, one cannot neglect the demand for improved urban mobility. This must be characterized by better access to opportunities, affordable and efficient public transport systems at the core of improving urban public transport infrastructure (Boschetti et al., 2014).

The existing body of knowledge reveals that the four interdependent main themes which serve as performance indicators of the public transport systems are accessibility, affordability, efficiency, and reliability or convenience (Diana & Daraio, 2010). Urban public transport systems that reflect a balance across these four main themes can meet the minimum requirements for improved urban mobility through an integrated urban public transport network. Subsequently, innovative urban public transport systems policy interventions should strive to achieve these main principles to ensure fully functional and economically viable cities.

3.5 South African Urban Public Transport Systems: An Overview

The South African Urban Public Transport is characterized by poor coordination of the multiple stakeholders and service providers responsible for public transport services for commuters (Patel et al., 2001). These range from formalized municipal buses in major cities, private owned bus companies, rail transport, and non-regulated, informal public transport in the form of taxis which transport most of the country's passengers in urban areas. Charma et al. (2017) observed that the apartheid system of governance has contributed to the current land-use pattern which perpetuates long travel distances, with the location of townships far from major economic nodes. This situation arose through racial and spatial segregation policies which led to fragmented spatial forms through provision of housing opportunities for the majority of the population far from major economic nodes, employment opportunities, and urban amenities.

The previously disadvantaged (Black people and people of color) were the main victims of such segregation policies and are to this day mostly reliant on public transport. Thus, investment in public transport infrastructure is central to spatial and socio-economic transformation. Maphakela et al. (2013: 200) noted that "the transportation burden faced by developing rural and urban communities on a daily basis in South Africa is real and substantial." It was noted that massive transport infrastructure investments have been channeled toward transforming the spatial form and socio-economic context. Nevertheless, slow progress on the ground leaves much to be desired regarding the effectiveness of policy and legislative interventions that have been employed, post-1994.

Emeran et al. (2013) further notes that, given the South African context, a public transport network of high-volume corridors should be a priority, with the aim of shortening the time for travel trips and improving access to economic opportunities. Of late, the South African national government has advocated the Integrated Mass Rapid Public Transport Networks (IRPTN) public transport infrastructure (van Ryneveld, 2008). Nevertheless, despite such an already established cause, Luke and Heyns (2013) observe that the country's achievements in urban public transport planning have been disappointing. This is despite good policy instruments and legislation informed by good principles which are based on the 1996 white paper which was aligned with international standards on public transport infrastructure.

3.6 Modes of Public Transport in South Africa

The South African public transport system is renowned for its distinct formal and informal public transport operators (Venter, 2013). Formal public transport constitutes the government-subsidized public system in the form of buses and trains, while the informal public transport network is predominantly the taxi industry, recognized by government but is not a subsidized operation (Williams & Kingma, 2002). Since the emergence of innovative approaches to urban public transport systems across the globe, Bus Rapid Transit (BRT) systems have gained momentum and are being implemented across South Africa. However, the technologically advanced train (Gautrain) is only limited to Gauteng as a provincial government-driven initiative. The mini-taxi industry and the Metrorail trains remain as transporters of the people from isolated townships, where the majority of the people reside.

Dawood and Mokonyama (2015) observed that the government subsidy has only been offered for the bus system and rail network. This excludes the taxi industry, arguably the biggest transporter of the urban poor segment of the population, and has also contributed to the lack of affordability of public transport by them. Further, limited access to economic opportunities by the urban poor due to the geographic location of some residents means that they are far from workplaces. As a result, the cost for transport increases, and they use more than one mode of public transport for a single trip (Franklin, 2014). This results in increased traveling times, which negatively affects productivity in the workplace and discourages some residents from working; some even give up their jobs (Mackie et al., 2012).

Notably, the government has little say over the taxi industry, responsible for transporting at least 65% of the public commuters in South Africa (Ahmed, 2004). Government initially attempted to formalize and recapitalize the mini-bus taxi industry, and later followed this by introducing innovative urban public transport systems (Fourie, 2003). However, it is important to note that the mini-bus taxi industry has played a critical role in the country's public transport arena, especially where formal public transport operations like buses and trains could not cater to the people's needs (Woolf & Joubert, 2009). However, there were also many inadequacies, with the taxi industry operations often associated with violence. Hence the need has been identified to create a safer and more convenient public transport system through an innovative urban network (Venter, 2013).

3.7 Public Transport and Socio-economic Transformation

Pre-1994, the South African economy was a polarized labor market, where benefits of most skilled jobs were accessible mainly to the minority (whites), while African people occupied low-income jobs in the unskilled or semi-skilled labor market (Leibbrandt et al., 2010). In addition to African people (being in a majority), and mostly belonging to the low-income segment of the polarized labor market, they were also located far away from economic opportunities through apartheid segregation laws. This meant that they had to travel long hours to work, with the most common mode of public transport being the mini-bus or Metrorail trains. Thus, it was not surprising that spatial and socio-economic transformation became South Africa's top priority post-1994 through policy directives and legislative frameworks (Seekings, 2014).

Nevertheless, most of the said policy pronunciation still seems to have a limited impact, given the prevailing circumstances of spatial segregation, with limited affordable public transport available for low-income earners who are reliant on it. Spatial structures and forms inherited from the apartheid planning system have hindered and frustrated social and economic transformation in most parts of the country (Turok, 2013). The income disparities in South Africa continue to reflect the historical imbalances manifested through a racial footprint and geographic location (Triegaardt, 2006). Evidently, there has been little or limited impact of the transformation strategies which are crucial for spatial integration, socio-economic transformation, and land reform, despite having a proliferation of micro-financing and institutional transformation (Baruah, 2009).

Travel costs and a lack of access to economic opportunities have been the predominant challenges, with the urban poor having to spend more than a third of their income on transport. They sometimes use more than one mode of public transport due to lack of modal integration, with a single travel fare per trip instead of paying for each mode of transport (Patel et al., 2001). The fragmented spatial form is central to the socio-economic challenges experienced by many, and there is now a renewed call for redressing the past spatial imbalances through transit-oriented development (National Planning Commission, 2011). The National Development Plan by the NPC clearly spells out that integration encompasses a number of factors and sectors that will result in best practices and will contribute to economic transformation when facilitated and coordinated correctly.

3.8 Gauteng Public Transport Infrastructure Policy Directives

In South Africa, national, provincial, and local governments all fund public transport infrastructure projects in line with their competency areas and core mandates. The Gauteng 25-year Integrated Transport Master Plan (ITMP 25) maps out policy directives and the desired future of the Gauteng public transport network that will respond to the challenges of spatial segregation and socio-economic disparities (Gauteng Government, 2013). The Gauteng ITMP 25 was derived from all existing transport plans across the region, and the Gauteng Department of Roads and Transport was tasked with the responsibility of managing and coordinating the implementation processes of the said plan.

The Gauteng Provincial Government has public transport projects that are driven from the provincial office, e.g., the Gautrain Rapid Rail. In addition, municipalities within the Gauteng province align their transport plan to provincial policy directives pronounced in the 25-year plan when executing their core functions within their area of jurisdiction. In this regard, the ITMP identifies and anchors four main areas of intervention: Land-Use Development; Strategic Public Transport Network; Freight transport; and Road Transport. These are discussed in detail in the ensuing sections.

(a) **Land-use development**

The Land-Use Development component articulates the need to promote densification along public transport corridors, which should incorporate subsidized housing options for the urban poor within the core urban areas. Efforts however remain questionable to get the private sector (developers) buy-in for the provision of inclusionary housing along densification corridors since they are the main role players in this regard. Also of significance is the call for public transport infrastructure investments to facilitate business growth and employment creation by boosting the local economy within and outside the urban core. However, the role of connector points and major stations as hotbeds for socio-economic innovation has hardly been explored, as there is no clear directive for the designing of stations or connector points.

(b) **Strategic public transport network**

The Strategic Public Transport Network policy announcement advocates for revitalization of the passenger rail network as being essential for spatial integration. The planner further makes provision for integrated rapid and road-based public transport networks as another public transport option for spatial and socio-economic transformation. Of paramount importance is the call for capacity building in the public transport sector. However, the less elaborated crucial aspect is the integration of the taxi industry (most commonly used by the previously disadvantaged) and other modes of transport to form a multi-modal public transport network.

(c) **Freight transport**

The core of ITMP strategies for freight transport is the need for intermodal freight hubs to transport goods. This is central to strengthening the local economy within the region and enabling easy imports and exports of goods through viable freight hubs in the region. Rail terminals for containers and consumer goods have been identified as areas with limited capacity, yet great potential, with the inland, freights being said to be constantly increasing. A freight ring road covering the whole Gauteng City Region linked to rail has also been advocated.

(d) **Road transport**

At the core of Gauteng's road transport policy directive is the call for a provincial-wide, public transport network that anchors improved urban mobility by providing efficient public options to enhance access to available opportunities. Managing travel demand, and thereby minimizing traffic congestion on public roads has been identified as essential for improving urban mobility. The need for promotion of Non-Motorized Transport has also been identified as one of the key strategies to get people off the road through the creation of mixed-use zones and activity corridors. This will empower people to live and work within the same precinct while having easy access to social facilities and economic opportunities through non-motorized transport.

3.9 Public Transport Infrastructure and Socio-economic Transformation in Gauteng

Given the socio-economic transformation policy directives, the Gauteng provincial government has outlined its plans to transform the economic space of Gauteng in the next 10–15 years (Makhura, 2015). This includes, for example, the multi-pillar policy directive on radical transformation, modernization, and reindustrialization geared toward social cohesion and economic inclusion. Public transport infrastructure investments form the core of this multi-pillar policy directive. Taking note of the aforementioned, it is nevertheless important that questions are probed, and new innovative approaches are researched and adopted to prevent repetition of major policy interventions with limited socio-economic transformation benefits and spatial integration.

In 2014, the Gauteng City Region Observatory (GCRO) produced a report on urban mobility highlighting key transport infrastructure investments and major transport routes in the Gauteng City Region (Wray & Gotz, 2014). However, the report did not adequately discuss the impact of innovative urban public transport systems implementation on socio-economic transformation through improved urban mobility, business growth, and employment creation. Rather it focused on implemented and planned projects, with no clear direction on what has been done to ensure that the implementation of such plans contributes to the abovementioned transformation. Musakwa and Gumbo (2017) also observed that the already implemented plans have yielded no positive results, but instead have perpetuated the fragmented spatial form and further deepened socio-economic challenges.

Despite the Gauteng ITMP calling for municipalities to work toward achieving the policy goals set out in the ITMP, there is a very limited collaborative approach to planning and implementation between the province and Municipalities. Even the IRPTN aligned innovative Gautrain, implemented across all three Metropolitan cities in Gauteng province and BRT systems suggest that there is limited collaboration between municipalities and provincial government. This is evidenced by the lack of multi-modal connector points and separate billing systems. Thus, provincial and municipal public infrastructure plans/investments have minimal operational and functional integration for convenient multi-modal public transport.

3.10 City of Tshwane Public Transport Services

The City of Tshwane urban public transport network is characterized by a combination of both road and rail public transport infrastructure. Road public transport operations in the city consist of a privately owned, a vibrant taxi industry, privately-owned bus systems, Municipal Metro Buses (Tshwane Bus Services), and the A Re Yeng BRT system. The Metrorail system consists, on the other hand, of Mass Rapid Transit transport managed by the National Department of Transport Agency, i.e., the Passenger Rail Agency of South Africa (PRASA). Of late, the innovative, technologically advanced rapid rail system has been established in the form of the Gautrain managed by the Gautrain Management Agency (2013) in conjunction with the Gauteng Department of Roads and Transport.

The taxi industry and Metrorail trains transport most of the Tshwane population, with the bus system (both Municipal buses and privately owned) following thereafter. However, the Gautrain and the A Re Yeng did not make a priority of those in dire need of public transport. This was because their operations mostly started in the main economic nodes and did not necessarily link isolated, overpopulated townships with the main economic nodes where business and employment opportunities are found. Therefore, it is not surprising that Metrorail trains and the mini-bus taxi industry are prominent and popular in the areas (townships) where the previous disadvantaged reside. The Metro buses run mostly in urban suburbs and a few sections of the black townships. The privately-owned buses also operate both in urban suburbs and some sections of the townships.

The Gautrain and the A Re Yeng operations started around Pretoria Central and Hatfield (with Gautrain services also available in the Centurion node), and plans are to expand the services to the townships and suburbs where they have not reached. The Gautrain links three major centers in the City of Tshwane namely: Centurion, Pretoria Central Business District (Bosman Station), and Hatfield. On the other hand, the A Re Yeng connects the Pretoria Central Business District and Hatfield Business node, with seven connector points or stations on the BRT route between Pretoria Central and Hatfield nodes. Thus, obvious calls have been made for an integrated public transport network within Gauteng as a policy directive by provincial government. Despite this, even the latest public transport investments in the City of Tshwane do not do justice toward attaining operationally integrated public transport operations for spatial integration and socio-economic transformation.

3.10.1 City of Tshwane Innovative Urban Public Transport Systems

Tshwane is among the 12 cities identified under the National Department of Transport public transport strategy for its Integrated Rapid Public Transport Network (IRPTN) initiative (Engineering News, 2012). The IRPTN initiative seeks to improve various modes of public transport consisting of commuter rail, Bus Rapid Transit (BRT), Bus services, Mini-bus Taxis, and Non-motorized transport. Since the inception of the IRPTN program, the said 12 identified cities are receiving massive investments from National government through the Public Transport Infrastructure and Systems Grant (PTISG) which municipalities must apply for when they roll out IRPTN. Tshwane serves as South Africa's administrative capital and is among Gauteng's three metropolitan cities; thus, the city has a significant role to play in shaping the direction of growth and development in the province and the country (City of Tshwane, 2016).

However, the accelerated modal upgrade initiative has slowly subsided, with minimal positive outcomes, and has been overtaken. The Integrated Rapid Public Transport Network (IRPTN) initiative seeks to produce a competitive and integrated urban public transport network. Within the City of Tshwane, the A Re Yeng Bus Rapid Transit is the only Municipal driven IRPTN aligned project which incorporates aspects of the latest innovations in urban public transport systems. However, it is not operationally integrated with other modes of public transport found in Tshwane, i.e., Mini-bus taxis, Metrorail mass rapid trains, private operated buses, or even the municipal operated Tshwane Bus Services (TBS). Rather, it has positioned itself as one of the public transport providers, with the establishment of a specialized unit located in Hatfield focusing purely on A Re Yeng Operations.

Not much can be said about the A Re Yeng contribution toward an integrated (multi-modal) urban public transport network consistent with the IRPTN national strategy. A Re Yeng is not even operationally integrated with Gautrain, which is another IRPTN aligned initiative. Within Tshwane, the Gautrain links three major centers in the City of Tshwane, namely Centurion, Tshwane Central Business District (Bosman Station), and Hatfield, with a future expansion planned for the Mamelodi township. However, even where there are stations of both BRT (A Re Yeng) and Gautrain, i.e., Hatfield and Pretoria central, there are no signs of precinct designs to enhance multi-modal integration. Thus, socio-economic benefits of these investments in strengthening the local economy leave much to be desired.

3.10.2 A Re Yeng BRT Network

According to the City of Tshwane, IDP (2016), the objectives of the City of Tshwane Integrated Rapid Public Transport Network (IRPTN) are to provide a high-quality, safe, and affordable public transport system. In 2011, the City of Tshwane Council approved implementation of Phase 1 recommendations for the innovative urban public transport systems as part of the IRPTN initiative (City of Tshwane, 2016). The said Phase 1 network is comprised of two main trunk line services and numerous feeder services supplemented by plans for improving the non-motorized transport network within the City of Tshwane. The two main trunk lines are: Line 1:

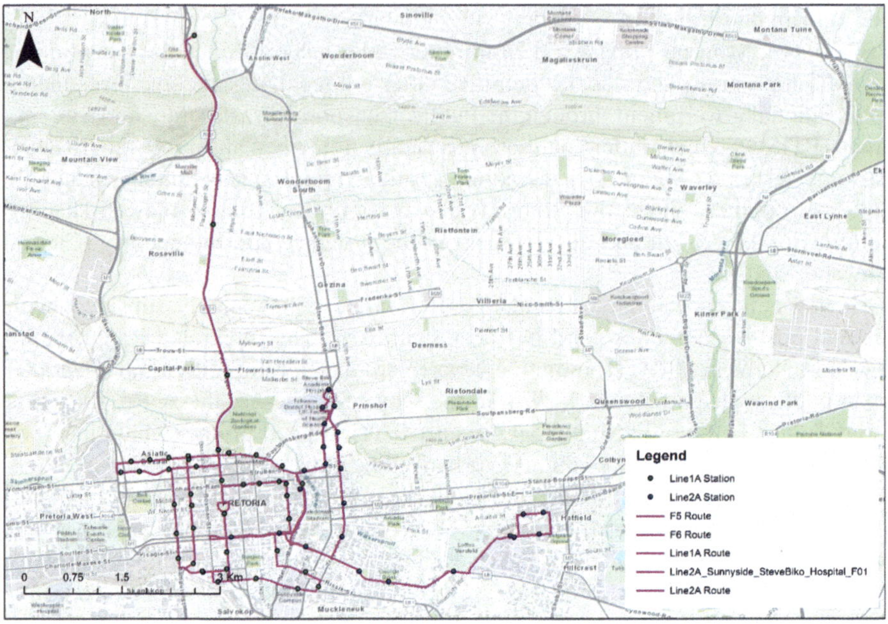

Fig. 3.1 A Re Yeng BRT network

Pretoria Central to Soshanguve (Kopanong) and Line 2: Pretoria Central to Mamelodi (Mahube Valley) (see Fig. 3.1).

Currently, the BRT network runs between the Pretoria CBD and Hatfield economic nodes, with seven (7) connector points in between. Though operations started around the City Center, plans are in place to expand the services to be fully operational and effective in townships and suburbs, where it has not reached. Future expansions for the A Re Yeng are envisaged to cover the wider City of Tshwane Municipal area of jurisdiction. However, operational isolation of the service and stations or connector points in relation to other modes of public transport falls short in terms of whether the current investments are contributing meaningfully to improving urban mobility and socio-economic transformation.

The said BRT implementation plans (completed and envisaged) lack a crucial element of modal integration with other modes of public transport such as the vibrant taxi industry that is privately owned, the bus systems that are privately owned, and Metro Buses in the form of Tshwane Bus Services. Thus, the introduction of innovative urban public transport systems that receive massive investments from government is yet to reach those in dire need of public transport which makes urgent the prioritization of public transport investments. These innovative transport systems have been introduced amid a fragmented public transport network, where various public transport providers operate independently from each other. Prevailing circumstances suggest that innovative approaches within the City of Tshwane seem to put less emphasis on stakeholder collaboration for the integrated/multi-modal public transport network.

3.11 City of Johannesburg Public Transport Services

The City of Johannesburg urban public transport network is characterized by a combination of both road and rail public transport infrastructure. Road public transport operations in the city consist of a vibrant taxi industry that is privately-owned, privately-owned bus operations, Metro Buses, and the Reya Vaya BRT system. There is also a strong presence of Metrorail and Gautrain services within the City of Johannesburg, which also connects commuters to the other two metros of the Gauteng province. The City of Johannesburg, commonly known as the economic hub of South Africa, and its most populated city is renowned for its strong presence of mini-bus taxi operators and popular use of private vehicles for daily commuting needs. It is not surprising that most Public Transport investment initiatives often prioritize interventions in the buzzing economic hub of South Africa. Even the BRT system was first introduced in this City before it moved to other metropolitan cities in the Country.

3.11.1 Innovative Urban Public Transport Systems in the City of Johannesburg

The Re Ya Vaya BRT bus service is one of the most advanced BRT systems in the country, covering a number of areas in the city, and was one of the few public transport systems that prioritized its main township base (SOWETO) where public transport is needed the most. While the Gautrain Rapid Train Service is another form of Innovative Public Transport Systems operational in the Metro, the Re Ya Vaya remains the only form of innovative public transport imitative fully management by the city. Of particular note, Johannesburg also has a strong focus on Corridor Development within its Transit-Oriented Development Program through its program called Corridors of Freedom. Notably, this initiative looks beyond providing innovative public transport options, but rather puts an emphasis on the need for sustainable land-use development and planning along with major public transport initiatives.

The corridors of freedom concept can be seen as a step closer toward integrating Non-Motorized Transport as part of its innovations within Public Transport Systems Planning. The Corridors of Freedom concept emphasizes creation of activity nodes at connector points and densification along public transport corridors with the objective of getting people, to live, play, and work or have access to economic activities in the same area. The City of Johannesburg is evidently making the necessary effort to ensure that there is a strong link between public transport planning, corridor development, and mixed-use development. However, the city has had minimal tangible initiatives toward multi-modal public transport operations integration.

3.12 City of Ekurhuleni Public Transport Services

The City of Ekurhuleni urban public transport network is characterized by a combination of both road and rail public transport infrastructure. Road public transport operations in the city consist of a vibrant taxi industry that is privately owned, bus systems that are privately owned, the Harambe BRT system, and City of Ekurhuleni Municipal Bus Services. The City of Ekurhuleni has an established unit responsible for all policy and strategic frameworks concerning all Public Transport Systems related operations so as to ensure customer satisfaction and optimization of public transport services in collaboration with public transport operators. Central to the Ekurhuleni public transport operations is enhanced urban mobility and effective and efficient functionality of towns and communities within the metro.

Notably, the City of Ekurhuleni is also home to an International Public Transport Node, namely, the OR Tambo International Airport. This facility is of international significance as it connects the local and international community with Gauteng, South Africa, and Southern African countries as well. It is therefore not surprising that the Ekurhuleni Metropolitan City embarked on the journey to brand itself as an aerotropolis, an initiative anchored through a 30 years long multi-billion Rands funding model linked to a Municipal Council approved development Plan. This initiative seeks to transform the Metro from being a shadow of the City of Johannesburg and bring about initiatives that will drive spatial and socio-economic transformation to the benefit of local communities, while positively contributing to the country's GDP.

3.12.1 Innovative Urban Public Transport in Ekurhuleni Municipal Area

The City of Ekurhuleni has unit responsibility for planning, coordinating, implementing and monitoring all Integrated Public Transport Network initiatives within the city. This unit anchors innovative urban public transport initiatives where the Harambe BRT system is managed. Through initiatives driven in this division or unit, the City of Ekurhuleni ensures that adequate public transport infrastructure is provided and that non-motorized transport and intelligent public transport systems are prioritized. The Harambe BRT system remains the symbol of Innovative Urban Public Transport Systems currently operated by the Metro, though there is also the Gautrain Rapid Rail Service operated by the Gauteng provincial government which has a strong footprint in the Metro, especially connecting the OR Tambo International Airport with the rest of Gauteng Province. Nevertheless, minimal effort toward creation of an operationally integrated multi-modal public transport network has been witnessed.

3.13 Significance of Collaborative Planning in Policy Formulation and Implementation

The complex challenges that development practitioners and urban planning professionals are confronted with every day require an integrated approach to development (Alfaro, 2014). In this regard, Healey (2003) made valuable contributions in setting the tone for key considerations when planning for fragmented communities and addressing socio-economic challenges through collaborative planning. Thus, the interdependent nature of the fragmented spatial form disintegrated urban public transport systems, the dire needs of South Africa, and persistent socio-economic challenges, as already argued in previous chapters, necessitated that this study adopts the collaborative planning theory.

The Collaborative Planning Process (CPP) was first introduced in the mid-80s through the book titled 'Collaborative Planning' (Civic Healey, 2003; Innes, 2015; Shakeri, 2011; Voice, 2015). This book and the subsequent discussion and publications seek to give a common goal to different sectors which work in isolation, resulting in duplication of responsibility and misuse of resources. The idea of collaborative planning draws all kinds of spatial, regional, and urban planning systems toward comprehensive and well-coordinated public sector planning, implementation, and overall operations and monitoring (Ndwandwe & Gumbo, 2017). Stakeholder engagement, collaborative efforts, and effective communication among interdependent institutions with overlapping functional areas and cross-cutting development issues are crucial to the Collaborative Planning Process, in order to effectively address socio-economic challenges and development needs.

According to Capello (2011), theories of planning and approaches to economic growth, development planning, and socio-economic transformation should be responsive to challenges in the day-to-day life of the residents in a city. Collaborative Planning is the most prominent theory of planning and arguably the most appropriate in dealing with complex development planning issues. This is because it identifies the key principles to be addressed and role players essential to the achievement of any set goals in the planning and development of towns (Kobler, 2009). Therefore, the Collaborative Planning Process allows for interdependent sectors of the economy, government departments and institutions to combine their efforts to address the needs of the community while ensuring that public funds are used effectively thereby contributing to socio-economic transformation.

3.14 Stakeholder Engagement and Effective Communication

Collaborative Planning has evolved over the years with a view to integrating various planning methods and ensuring effective planning which achieves desired outcomes within a favorable timeframe (Brand & Gaffikin, 2007). Notably, political will and shared vision by all stakeholders are of paramount importance as the first building block to enable fruitful collaborative efforts. Moreover, collaborative planning is a communicative and interactive form of planning practice that is dependent on clear guidelines toward favorable outcomes (Shakeri, 2011). Without effective communication to achieve consensus, it is impossible to realize the fruits of collaborative planning, therefore, planner(s) should facilitate the process in such a way that informed decisions are made while also achieving consensus among stakeholders (Sukol, 2012). Furthermore, communication mechanisms and channels must be established among stakeholders to attain favorable outcomes.

Planning practices have however evolved, as has the role of the planner in shaping the spatial form and growth or development pattern of a city or town (Kobler, 2009). Collaborative planning means that the planner has become more of a facilitator with technical input, as opposed to a technical expert entrusted with the responsibility of attending to community development challenges (Peterman, 2004). This is even more true, considering that members of the public should form the core contributors to the growth and development of a city or town. According to Aapaoja et al. (2013), at the inception phase of any development, relevant stakeholders must be identified, engaged, and involved throughout the process so that every stakeholder's input is considered to create a collective strategy.

Also, as noted by Baptista (2005), collaborative planning gives meaning to everyday life, where the prevailing circumstances and contentious issues in the society are not left unattended. As a result, a collaborative planning process is not the opposite of conflict but rather, leads to tensions which create a platform for the stakeholders to learn about each other. More options can thus be generated through proper conflict management by the planner as a technical facilitator, consensus can be achieved, and everyone will pull in the same direction (Innes, 2015). Thus, a well-coordinated stakeholder engagement, together with improved citizen participation is essential for building a common ground and ensuring everyone's buy-in while addressing cross-cutting issues. In the end, therefore, the process should be associated with planning with the people as opposed to planning for the people.

3.15 Effectiveness of Collaborative Planning Process in Practice

Comprehensive outcomes of the Collaborative Planning Process cannot be achieved without good governance at the center of stakeholders' coordination. Kobler (2009) distinguished government from governance in the sense that government is about provision of services, while good governance is the ability to lead a society with different interests and get them to work toward a common goal. Thus, it is essential that the planner (as a facilitator with technical input) ensures that there is buy-in from all stakeholders and that everyone is working toward a common goal. In spite of its importance, there has been limited examination of the impact of the Collaborative Planning Process (Laurian & Shaw, 2009). Thus, it must be noted that, since the introduction of collaborative planning in the 1980s, scholars have studied how guidelines for collaborative planning have helped in dealing with day-to-day challenges within society.

For example, scholars such as Frame et al. (2004), Susskind et al. (2010), Wondolleck and Yaffee (2000), and Moote et al. (1997), are among those who have contributed effectively to assess the applicability of the collaborative approach to planning in real situations. The common positives, as observed by these scholars include innovative solutions of high-quality; fair representation of all interested and affected parties; collaborative effort and teamwork toward a common goal; increased investment capital; optimal utilization of resources, and an innovative shared knowledge base. However, there are challenges associated with the Collaborative Planning Process that should be guarded against. These include level of influence of different stakeholders; increased and continued conflicts; loss of interest by some stakeholders and under-representation of some stakeholders.

Collaborative Planning Process principles are central to determining a level of collaboration among interdependent stakeholders through integration of various dysfunctional land uses inherited from the colonial and apartheid era. Provision of affordable and efficient public transport systems for the previously disadvantaged becomes central to the development of a collective strategy for improving urban mobility while contributing to socio-economic transformation. In synthesizing key indicators for the effectiveness of collaborative planning in the context of this study, six central components are identified which determine the usefulness of the Collaborative Planning Process to redress socio-economic challenges.

3.16 Core Components of Collaborative Planning Process

Observations and discussions by scholars like Healey (2003), Kobler (2009), Shakeri (2011), and Sukol (2012) form the cornerstone of collaborative planning. The above-mentioned core components can be summarized as follows: Pre-existing conditions/status quo; Stakeholder identification and engagement; Level of inclusiveness;

Level of authority; Communication and conflict management; and Formulation of options for stakeholders. These are discussed in detail below.

(a) **Prevailing circumstances (Pre-existing conditions/status quo analysis)**

Though it is important for cities and towns to learn from each other and create a platform for sharing of ideas, this must be done with an understanding that places are unique and should be treated as such for the development processes to ensure quality of life in a city or town (Kavaratzis & Ashworth, 2005). Thus, there is a need to understand prevailing circumstances of any given area, from which adequate planning programs and processes can be developed without neglecting local knowledge. Accordingly, a diagnostic of local development and socio-economic challenges for understanding local issues and local context-based research knowledge should form the basis for the planning processes. This should then inform the rest of the process and ensure formulation of well-researched policy instruments that respond to local development needs and challenges, where public investments can be channeled appropriately.

The level of access to basic amenities, resources, and socio-economic opportunities should form the basis for investment priorities. Such should be established during the process of conducting status quo analysis as part of the research and local knowledge development. The complex nature of the socio-economic needs of the community should be carefully studied to ensure that government investments and policies do not only focus on infrastructure development and basic amenities but also incorporate the identification of necessary trickle-down effects in the form of socio-economic benefits.

(b) **Stakeholder identification and engagement**

Baptista (2005) argues that no single institution can effectively attend to the development and socio-economic needs of the people, hence the need for stakeholders to acknowledge and embrace their interdependent nature. Consequently, a status quo report is central for the identification of interdependent sectors and the role players which are important in addressing challenges identified. Stakeholder identification should be directly linked to all aspects of the development project or policy such that community development is effectively facilitated to achieve the aspirations of the people. Governments' organizational structure(s) categorize provision of different public services as core mandates of different departments. Even so, the interconnected nature of some departments' core functions is such that prominent success is impossible except through a collective strategy for investment programs and policy directives.

During stakeholders' engagement, all the stakeholders must feel some sense of responsibility for the project and take ownership to achieve a desired spatial form. Some stakeholders will eventually become responsible for funding some aspects of a capital investment framework for a collective strategy. As such, government department investments can then be easily channeled toward holistic development programs that will effectively transform people's lives. Public–Private Partnerships

(PPPs) should also be championed, with the state being an enabler by allowing business growth through government-led development initiatives which comprehensively address socio-economic challenges.

(c) **Level of inclusiveness**

The Civic Voice (2015) advocates for a clear distinction between stakeholder engagement through public participation, and consultation of already formulated development proposals. As argued in some of the sections above, improved citizen participation is central for inclusiveness of the state in policy and plan formulation and decision making. All stakeholders affected by the implementation of the envisaged development programs, either as investors, project funders, or beneficiaries should form part of every decision-making process, such that everyone can take full ownership of the development program (Baptista, 2005). The interactions must therefore be affected at all levels, with all stakeholders having an opportunity to be engaged and make meaningful contributions. Community involvement during formulation of development proposals should also be a priority. Kobler (2009) observes that planners are instrumental as facilitators between citizens and government to build community capacity. Platforms for robust debates must be created for synergies during policy formulation through scenario planning under the Collaborative Planning Process to allow for a focused, engaged course of action. This is consistent with Shakeri (2011), who argues for a focused inclusive process with effective interaction, where citizens are given enlightenment and better understanding of all aspects of development, i.e., the effectiveness of the process as well as implementation of collaborative planning outcomes. Improved participation will also help in ensuring responsible citizenship and limit vandalism of the infrastructure, as the community takes ownership of the whole project.

(d) **Level of authority**

Healey (2003) argues that the Collaborative Planning Process should not be mistaken as neutralization of the authority of those entrusted with the responsibility to drive government service delivery. Rather, it should be seen as a platform, where all stakeholders form the core of the decision-making structure, compared to the consultation approach, where one stakeholder consults others and independently makes decisions. Notably, Baptista (2005) observes that decisions made independently by any stakeholder are not as effective as decisions which are outcomes of exchange of ideas by various stakeholders, where everyone takes ownership of development. In the same vein, the Civic Voice (2015) argues for inclusive development.

Collaborative Planning Processes create platforms for innovative approaches to be developed and eliminate centralization of decision making which may result in omission of some crucial aspects of development (Alfaro, 2014). A learning platform is created for everyone, as stakeholders begin to appreciate the interdependent nature of their core functions, while participatory processes help achieve social cohesion (Servaes, 2008). Innes (2015) argues that communication is critical in ensuring stability, as it helps stakeholders find a common ground over power relations, where levels of interaction are developed and challenged as necessary to ensure effectiveness. The state therefore effectively becomes an enabler, while stakeholders make meaningful contributions to policy and development decisions.

(e) **Communication and conflict management**

Under Collaborative Planning Processes, planning has evolved to become a communicative as opposed to a technical and scientific process, where technical experts like planners and engineers develop plans and consult the public from an already formulated development proposal. This results in conflicts and resistance from civic organizations and the public (Civic Voice, 2015). In contrast, Innes (2015) observes that some critics of the Collaborative Planning Process argue that peer pressure often takes precedence, with some aspects of theorizing about a good city being eliminated in the process. However, counterarguments are that platforms for new knowledge systems and information sharing are created through Collaborative Planning Processes, while conflict and mistrust between government and citizens can be redressed (Kobler, 2009).

Observations by Innes (2015) see conflict as a critical instrument in collaborative planning as more options can be derived from the process. In this regard, the role of a planner as a facilitator of the process is even more essential and critical. It is crucial to establish norms and standards among stakeholders to effectively communicate with each other and eliminate conflict (Baptista, 2005). This will enable all stakeholders to be actively involved in shaping the future of development directions or desired spatial form. This is what a Collaborative Planning Process within communicative planning theories is embedded in, i.e., creating a platform for well-informed decisions and desired outcomes.

(f) **Formulation policy options for stakeholders**

The formulated options must be based on the issues agreed upon and proposals made collectively during the stakeholder engagement process (Baptista, 2005). Once pre-existing conditions are established and understood by all stakeholders with insights on levels of access to basic amenities and resources, a collective strategy can be developed. The role of planner(s) as facilitators then becomes instrumental in formulating different policy options and investment directions for stakeholders. Development proposals should be responsive for development needs and challenges across decision-making areas, thus reflecting a synthesis of all possible development paths that can be pursued together with their consequences. The stakeholders explore various policy options and decide on the best possible collective strategy for investments.

(g) **Effectiveness of post-apartheid policy interventions**

Almost all development initiatives and investments in post-apartheid South Africa are centered on socio-economic transformation and spatial transformation. Hence the need exists for an integrated approach to innovative urban public transport systems and socio-economic transformation for spatial integration and creation of vibrant economic nodes. In actual fact, regardless of complex development challenges facing urban communities in South African cities, there are substantial investments being made. Those investments should make meaningful contributions to spatial integration and socio-economic transformation. However, coordination and collaboration among stakeholders is a concern because of the absence of a collective strategy with an implementation plan.

Government expenditures are "tick-the-box" exercises, where more often than not, practice will precede research resulting in the development programs being implemented without even understanding the local context. As an example, in the City of Tshwane, because of political pressure, government officials are driven by how much they spend from the allocated budget, without any measure being put in place to ensure holistic sustainability of government's investment program.

Many policies and programs have been put in place by the South African government toward social and economic transformation for sustainable livelihoods, but the prevailing circumstances leave much to be desired (Gumede, 2013). Statistics suggest that there has been little or no progress if any in bridging the inequality gap and poverty alleviation. Central to this is the ever-increasing problems of unemployment, labor market polarization, informal settlements low literacy skills, and lack of access to basic amenities in most rural communities of the country. This is observed, while the state continues to produce policies and plans that look good on paper, but which have limited or no significant impact in improving people's lives (Cilliers & Camp, 2013).

(h) **Prevailing circumstances**

There have been various attempts by the democratic government to improve the social and economic status of people and build sustainable communities, but situations on the ground suggest otherwise. These include the Reconstruction and Development Program (RDP); the 1996 Growth Employment and Redistribution (GEAR); Urban Renewal Program of 2000; Accelerated and Shared Growth Initiative—South Africa (ASGISA); the New Growth Path of 2010 and now the National Development Plan Vision 2030 (van Der Westhuizen, 2012). In the public dialogues, service delivery protests and economic disparities in the country have been the focal point of the discussion, where the inclusive nature of government policy initiatives has been questioned. The defining feature agreed upon in all these dialogues has been the notion that issues of economic disparities and labor market polarization persist and coexist.

The whole world may be working toward sustainable urban public transport systems, but the local circumstances that prevail in each country are unique, requiring local context-based policy interventions. In South Africa, Public transport interventions through a network of activity corridors are characterized by densification along these corridors. Activity nodes in major intersections thus present a great opportunity for mitigating the apartheid spatial imbalances. The end goal should be sustainable human settlements development, characterized by reduced travel times, affordable transport fares, and improved access to economic opportunities (Patel et al., 2001). Moreover, densification along activity corridors and mixed-use zones in major intersections will enhance spatial integration.

Unaffordable travel costs for the urban poor and limited access to economic opportunities due to their remote residential areas is a serious cause for concern. Despite mini-bus taxis being the most common public transport in these remote areas, government has prioritized bus services and rail networks and excluded the taxi industry (Dawood & Mokonyama, 2015). However, the fact that the taxi industry is unregulated cannot be used as a justification for not subsidizing taxis, as it is government's function to regulate public transport. Commuters should not pay unaffordable public transport fares due to government failure to regulate the taxi industry.

3.17 Lack of Practical Collaboration on Policy Implementation

Post-1994, South Africa has emphasized spatial and socio-economic transformation through policy pronunciations to deal with past spatial injustices to alleviate poverty, reduce unemployment, reduce the ever-widening inequality gap and support the economy. Nevertheless, apart from visible physical infrastructure, spatial integration and socio-economic transformation as realistic indicators of transformative policy and legislative interventions have lacked. Policy frameworks have been good on paper but require a practical collective implementation strategy. However, the autonomous approach to implementation by each sphere of government or even sector departments has meant that all public transport infrastructure investments consistent with policy and legislative frameworks have indirectly perpetuated spatial and socio-economic disparities.

Thus, though commendable work has been done through innovative urban public transport systems, the criteria for prioritization of policy interventions must be reviewed, as those who need public transport the most appear to be far from being a top priority. Lack of cooperation and coordination between the three spheres of government was found to be a hindrance to effectiveness of public transport infrastructure investments. It is therefore evident that all spheres of government and some sector departments are interdependent in their core functions, despite each having its own constitutional responsibility and legislative autonomy. Thus, a collective implementation strategy is central to redressing spatial and socio-economic disparities.

3.18 Conclusion: Toward Collective Strategy Formulation

Evidently, there is a need for more innovative approaches to collective strategy formulation by interdependent spheres of government and sector departments on policy implementation. They need to carefully understand the interdependent nature of their core functions in relation to public transport market trends in order to attain holistic policy interventions characterized by qualities of integrated spatial form and socio-economic transformation. The Collaborative Planning Process represents a crucial missing component in the current approach to policy implementation by spheres of government and sector departments. Thus, it is advocated as a tool that can be used for both policy formulation and a collective implementation strategy. Through collaborative planning, the effectiveness of intergovernmental investments will be evident through integrated spatial form and socio-economic prosperity.

Nevertheless, the Collaborative Planning Process is such that it must be robust and facilitated by well-trained urban planners to attain spatial and socio-economic benefits. Healey (2003) argues against the general misconception that a collaborative planning process is nothing beyond a means to an end. Healey (2003) advocates for an inclusive and interactive way of thinking, where stakeholders contribute meaningfully toward a mutually beneficial collective implementation strategy. The role of planners as facilitators of the process must be to ensure that contributions by all relevant role players are considered and incorporated in the final proposals, from the inception phase, until the scenario planning to develop policy options toward a collective implementation strategy.

Where trade-offs are made, analysis of arriving at the trade-offs must be presented as part of the development proposal options to the stakeholders. All deliberations and contributions made can either be accepted or rejected by stakeholders, depending on the validity and feasibility of development proposals put before stakeholders. Shakeri (2011) also notes that communication platforms must be created for deliberation among stakeholders to ensure that all contributions are factored into a collective implementation strategy. The public sector should initiate an intergovernmental research unit for public transport infrastructure. The purpose of this must be to influence the collective of stakeholders toward an effective implementation strategy that is responsive to local pre-existing conditions and development needs and challenges. This will enable identification of state departments that need to combine resources for their investments and complement each other toward spatial integration and socio-economic transformation.

References

Aapaoja, A., Haapasalo, H., & Söderström, P. (2013). Early stakeholder involvement in the project definition phase: Case renovation. *International Scholarly Research Notices, 2013.*

Ahmed, Y. (2004). Transformation of public transport operations from informal to formal services: An examination of initiatives by the Western Cape Provincial Department of Transport to transform the minibus-taxi industry.

Albalate, D., & Bel, G. (2009). *Factors explaining urban transport systems in large European cities: A cross-sectional approach.* Research Institute of Applied Economics, Working Papers 2009/05.

Alfaro, M. B. (2014). Collaborative decision-making processes in planning: Opportunities and challenges in the City of Rafaela, Argentina.

Aljoufie, M., Zuidgeest, M., Brussel, M., & van Maarseveen, M. (2011). Urban growth and transport: Understanding the spatial temporal relationship.

Anderson, S. P., & Wilson, W. W. (2014). Market power in transportation: Spatial equilibrium under Bertrand competition. *Economics of Transportation, 4*(1–2), 7–15.

Baptista, I. (2005). Is there a theory of collaborative planning we can talk about?

Baruah, B. (2009). Monitoring progress towards gender-equitable poverty alleviation: The tools of the trade. *Progress in Development Studies, 9*(3).

Boschetti, F., Maurizi, I., & Cré, I. (2014). *Innovative urban transport solutions*. Available from: http://civitas.eu/sites/default/files/civitas-plus-innovative-urban-transport-solutions-www-final.pdf. Accessed October 04, 2017.

Brand, R., & Gaffikin, F. (2007). Collaborative planning in a noncollaborative world. *Planning Theory, 6*(3), 282–313.

Capello, R. (2011). Location, regional growth and local development theories.

Cerna, L. (2013). The nature of policy change and implementation: A review of different theoretical approaches.

Charma, A., Tonkin, C., Denoon-Stevens, S., & Demeestére, R. (2017). Post-apartheid spatial inequality: Obstacles of land use management on township micro-enterprise formalisation.

Cilliers, J., & Camp, H. (2013). *Highway or byway? The National Development Plan 2030* (Vol. 6, pp. 1–16). African Futures Paper.

City of Tshwane. (2016). *Tabling of the city of Tshwane Integrated Development Plan for 2016–2021*. Available from: http://www.tshwane.gov.za/Documents/Tabling%20of%20the%20City%20of%20Tshwane%20Integrated%20Development%20Plan%20for%202016%202021.pdf. Accessed June 18, 2017.

Civic Voice. (2015). *Collaborative planning for all*. Available from: http://www.civicvoice.org.uk/uploads/files/Collaborative_planning_1.pdf/. Accessed October 04, 2021.

Clark, P. (2000). The influence of public transport on spatial development in Metropolitan Cape Town: Past, present and future.

Danish Business Authority. (2012). The future of eco-innovation: The role of business models in green transformation. In *OECD/European Commission/Nordic Innovation Joint Workshop*.

Dawood, G., & Mokonyama, M. (2015). *Towards a more optimal passenger transport system for South Africa: Design of public transport operating subsidies*. Council for Scientific and Industrial Research (CSIR).

De Den Hertog, J. (2010). Review of economic theories of regulation.

Department of Transport. (1996). *White paper on National Transport Policy*. Available from: https://www.comair.co.za/Media/Comair/files/level-playing-field/White-Paper-on-Nat-Transport-Policy-1996.pdf. Accessed October 05, 2017.

Diana, M., & Daraio, C. (2010). Performance indicators for urban public transport systems with a focus on transport policy effectiveness issues.

Dubina, I. N., Campbell, D. F., Carayannis, E. G., Chub, A. A., Grigoroudis, E., & Kozhevina, O. V. (2017). The balanced development of the spatial innovation and entrepreneurial ecosystem based on principles of the systems compromise: A conceptual framework. *Journal of the Knowledge Economy, 8*, 438–455.

Emeran, H., Sanders, S., Dyer, R., & Heyns, W. (2013). *PRASA: Delivering the National Strategic Plan*. Available from: https://repository.up.ac.za/bitstream/handle/2263/33246/Heyns_Prasa%20%282013%29.pdf?sequence=1. Accessed October 04, 2017.

Engineering News. (2012). Integrated rapid public transport network. *Engineering News*. Available from: www.engineeringnews.co.za. Accessed September 30, 2017.

eThekwini Municipality. (2016). Integrated transport plan.

Fourie, L. J. (2003). Rethinking the formalisation of the minibus-taxi industry in South Africa.

Frame, T. M., Gunton, T., & Day, J. C. (2004). The role collaboration in environmental management: An evaluation of land and resource planning in British Columbia. *Journal of Environmental Planning and Management, 47*(1), 59–82.

Franklin, S. (2014). Job search, transport costs and youth unemployment: Evidence from urban Ethiopia.
Fung, K. C., García-Herrero, A., Iizaka, H., & Siu, A. (2006). Hard or soft? Institutional reforms and infrastructure spending as determinants of foreign direct investment in China. *Documentos de Trabajo, 0616*.
Gakenheimer, R. (1999). Urban mobility in the developing world. *Transportation Research Part a: Policy and Practice, 33*(7–8), 671–689.
Gauteng Government. (2013). *Gauteng 25-year integrated transport master plan*. Gauteng Government. Available from: http://www.gauteng.gov.za/government/departments/roa/Documents/ITMP25-Main-Report-Final%20Nov13.pdf. Accessed November 14, 2017.
Gumede, V. (2013). African economic renaissance as a paradigm for Africa's socio-economic development. In K. Kondlo (Ed.), *Perspectives in thought leadership for Africa's renewal*. AISA Press.
Healey, P. (2003). Collaborative planning in perspective. *Planning Theory, 2*(2), 101–123.
Hook, W., & Howe, J. (2005). *Urban transport and the millennium development goals*. Available from: https://www.itdp.org/wp-content/uploads/2005/07/Transport-and-the-MDGs_ITDP.pdf. Accessed October 04, 2017.
Innes, J. (2015). A turning point for planning theory? Overcoming and dividing discourses. *Planning Theory, 14*(2), 195–213.
Jennings, G. (2015). Public transport interventions and transport justice in South Africa: A literature and policy review.
Jones, P., Turner, D., & Heydecker, B. (2014). Congestion management for China's transit metropolis cities supported by the UK Foreign and Commonwealth Office Prosperity Fund.
Karim, D. M. (2017). Creating an innovative mobility ecosystem for urban planning areas.
Kavaratzis, M., & Ashworth, G. J. (2005). City branding: An effective assertion of identity or a transitory marketing trick? *Journal of Economic Social Geography, 96*(5), 506–514.
Kobler, A. (2009). Building community capacity: How collaborative planning is changing the culture of governance in Seattle.
Laurian, N., & Shaw, M. M. (2009). Evaluation of public participation. *Journal of Planning Education and Research, 28*(3), 293–309.
Leibbrandt, M., Woolard, I., McEwen, H., & Koep, C. (2010). Better employment to reduce inequality further in South Africa. Tackling inequalities in Brazil, China, India and South Africa. The role of labour market and social policies. *OECD Library*.
Litman, T. (2012). *Comprehensive transport planning framework: Best practices for evaluating all options and impacts*. Victoria Transport Policy Institute. Available from: http://www.vtpi.org/comprehensive.pdf. Accessed March 10, 2015.
Litman, T. (2014). *Introduction to multi-modal transportation planning: Principles and practices*. Victoria Transport Policy Institute. Available from: http://www.vtpi.org/multimodal_planning.pdf. Accessed March 15, 2015.
Lucas, K., Tyler, S., & Christodoulou, G. (2009). Assessing the 'value' of new transport initiatives in deprived neighborhoods in the UK.
Luke, R., & Heyns, G. (2013). Public transport policy and performance: The results of a South African public opinion poll.
Mackie, P., Laird, J., & Johnson, D. (2012). *Buses and economic growth: Main report*. University of Leeds.
Makhura, D. (2015). *Gauteng state of the province speech 2015*. Gauteng Provincial Government.
Maphakela, W., Mashiri, M., Chakwizira, J., & Mpondo, B. (2013). *Building a sustainable platform for low-cost mobility in South Africa*.
Markovich, J., & Lucas, K. (2011). The social and distributional impacts of transport: A literature review.
Marrian. (2001). Towards a general theory of corridor development in South Africa.
Mazaza, M. (2002). The transportation/land use connection: A dilemma for the 21st century city?

Ministry of Transport. (2019). Annual report 2019/2020. https://www.transport.gov.za/documents/11623/41419/Final_DoT_Annual+Report_201920_02_11_2020.pdf/da1f3c68-3103-4df8-a67c-5532599d9acd
Moote, M. A., McClaran, M. P., & Chickering, D. K. (1997). Theory in practice: Applying participatory democracy theory to public land planning. *Environmental Management, 21*(6), 877–889.
Mulenga, G. (2013). *Developing economic corridors in Africa: Rationale for the participation of the African Development Bank.* NEPAD, Regional Integration and Trade Department.
Musakwa, W., & Gumbo, T. (2017). Impact of urban policy on public transportation in Gauteng, South Africa: Smart or dumb city systems is the question.
National Planning Commission. (2011). *National development plan vision for 2030.* https://www.gov.za/sites/www.gov.za/files/devplan_2.pdf. Accessed October 06, 2017.
Ndwandwe, B., & Gumbo, T. (2017). Exploring the efficacy of innovative urban public transport infrastructural systems on economic transformation: Case of Gautrain and Are Yeng in the City of Tshwane. In *Proceedings of the 4th International Conference on Infrastructure Development and Investment Strategies for Africa*, 30 August–1 September 2017, Livingstone, Zambia (pp. 130–146).
Page, J. H. (2012). *Legislative frameworks and policies.* Available from: http://dspace.nwu.ac.za/bitstream/handle/10394/8512/Page_JH_Chapter_3.pdf?sequence=4. Accessed October 04, 2012.
Pardo, F., Jiemian, Y., Hongyuan, Y., & Mohanty, C. R. (2010). *Shanghai manual—A guide for sustainable urban development in the 21st century.* Available from: https://sustainabledevelopment.un.org/index.php?page=view&type=400&nr=633&menu=1515. Accessed September 18, 2015.
Patel, H., Freeman, P. N. W., & Mitchell, M. F. (2001). Addressing the social aspects of urban transport through a more effective funding strategy in South Africa.
Peng, Z. R. (2005). Urban transportation strategies in Chinese cities and their impacts on the urban poor.
Peterman, M. (2004). Advocacy vs collaboration: Comparing inclusionary community planning models. *Community Development Journal, 39*(3), 266–276.
Polis, I. C. (2015). *The deployment of public transport innovation in European cities and regions.* www.polisnetworl.ew
Secure Technology Alliance. (2017). *Multimodal payments convergence—Part one: Emerging models and use cases.* Available from: https://www.securetechalliance.org/wp-content/uploads/Multimodal-Payments-Convergence-White-Paper-FINAL4-Mar-2017.pdf. Accessed October 04, 2017.
Seekings, J. (2014). *South Africa: Democracy, poverty and inclusive growth since 1994.* Available from: http://www.cde.org.za/wp-content/uploads/2014/04/democracy-works---south-africa-conference-paper---democracy-poverty-inclusive-growth-since-1994---by-jeremy-seekings-pdf-.pdf. Accessed October 05, 2017.
Servaes, J. (2008). *Communication for development and social change.* UNESCO.
Shakeri, M. (2011). Democratic assessment of collaborative planning practices.
Sodhro, A. H., Obaidat, M. S., Abbasi, Q. H., Pace, P., Pirbhulal, S., Fortino, G., Imran, M. A., & Qaraqe, M. (2019). Quality of service optimization in an IoT-driven intelligent transportation system. *IEEE Wireless Communications, 26*(6), 10–17.
Sukol, M. L. (2012). *Theorizing planning practice: Collaborative planning for smart growth on Long Island, New York.*
Susskind, L., van der Wansem, M., & Ciccareli, A. (2010). Mediating land use disputes in the United States: Pros and cons.
Triegaardt, J. D. (2006). Poverty and inequality in South Africa: Policy considerations in an emerging democracy.
Tsay, S., & Herrmann, V. (2013). Rethinking urban mobility: Sustainable policies for the century of the city.

Turok, I. (2013). The resilience of South African Cities a decade after local democracy. *Sage Journals: Environment and Planning, 46*(4), 749–769.
Van Der Westhuizen, C. (2012). South Africa: Economic growth, poverty and inequality. In *Foresight Africa: Top priorities for the continent in 2012* (pp. 33–34).
Van Ryneveld, P. (2008). *15 year review of public transport in South Africa with emphasis on metropolitan areas*. Hunter van Ryneveld (Pty) Ltd.
Vasconcellos, S. C. (2004). An integration proposal to the transport system of the city of Rio de Janeiro.
Venter, C. (2013). The lurch towards formalisation: Lessons from the implementation of BRT in Johannesburg, South Africa. *Research in Transportation Economics, 39*(1), 114–120.
Vermesan, O., & Friess, P. (2013). Internet of things: Converging technologies for smart environments and integrated ecosystems.
Vuchic, V. R. (2002). Urban public transport systems.
Wilkinson, P. (2006). Transit orientated development: A strategic Instrument for spatial restructuring and public transport system enhancement in South African cities.
Williams, R. M., & Kingma, R. (2002). Cape Town's strategic public transport network.
Wondolleck, J. M., & Yaffee, S. L. (2000). *Making collaborative work: Lessons from innovation in natural resource management*. Island Press.
Woolf , S. E., & Joubert, J. W. (2009). *A look at paratransit in South Africa*. Available from: http://www.up.ac.za/media/shared/404/ZP_Files/Innovate%2009/Articles/a-look-at-paratransit-in-south-africa_woolf-and-joubert-web.zp40153.pdf. October 05, 2017.
Wray, C. & Gotz, G. (2014). *Mobility in the Gauteng city-region*. Gauteng City-Region Observatory.
Wright, L. (2014). *Sustainable transport: A sourcebook for policy-makers in developing cities*. Available from: https://www.itdp.org/wp-content/uploads/2014/07/Module-3b-Bus-Rapid-Transit.pdf. Accessed October 18, 2020.
Zottis, S. (2014). Brazilian National Association of Transport Operators seminar examines response to rising social demand for urban transport. Available from: http://thecityfix.com/blog/brazilian-national-association-transport-operators-ntu-seminar-examines-rising-social-demand-urban-transport-brt-road-safety-qualionibus-toni-lindau-luisa-zottis/. Accessed October 04, 2020.

Part III
Physical Planning and integration of Urban Public Transport Systems

Chapter 4
Internet of Things and Urban Public Transport Systems in Johannesburg, Tshwane, Ekurhuleni, Cape Town, and Bulawayo

Abstract In the Fourth Industrial Revolution Era, cities of the developing countries, particularly in the African, Asian, and Latin American continents have experienced growing concerns about the integration of technology in public transportation systems. One of these concerns has been the lack of well-integrated, reliable, and efficient public transport systems. This lack exists mainly in urban centers due to rapid growth of the urban population, which has coincided with the end of colonialism, and which has given rise to large-scale economic, spatial and structural transformation of urban landscapes. Studies have shown that technological advances through the Internet of Things such as e-hailing services, e-ticketing, route planning, and trip planning can develop well-functioning innovative public transport systems. Therefore, this chapter unpacks how the Internet of Things can improve public transportation systems in the developing world cities in South Africa and Zimbabwe. Findings reveal that emerging technologies have been introduced in public transportation, but in order for the emerging public transportation systems such as the BRTs and RRTs to achieve the benefits of the Internet of things, more investment is required to ensure they are woven holistically into their operations and integrated with other public transportation systems.

Keywords Internet of things · Route networks · Urban public transport · Fourth industrial revolution · Metropolitan cities

4.1 Introduction

In contemporary research, there has been a rapid growth in efforts to understand the uses of web 2.0 big data. The Internet of Things (IoT) is an emerging research hotspot attracting international attention. It employs integration of various information detection equipment such as sensors, global positioning systems, laser scanners, and the Internet combined to form a vast network. Through this network, real-time collected information is transmitted to the end-user for the purpose of intelligent identification, deployment, monitoring, and management. Most smart devices are evolving and now have GPS technology integrated into them (Kemajou et al., 2019; Vecchio & Tricarico, 2019). Moreover, location-based services have advanced immensely, with

T. Gumbo et al., *Urban Public Transport Systems Innovation in the Fourth Industrial Revolution Era*, https://doi.org/10.1007/978-3-030-98717-6_4

the rapid investment in Web 2.0 platforms such as Strava Metro, Here, and Foursquare (Chai et al., 2019; Musakwa & Selala, 2016; Sun et al., 2017).

As Web 2.0 promises to be a new untapped data source, new research paradigms are currently being studied. These data have been utilized in transportation, with key focus on location and data velocity (Marzano et al., 2019; Vermicelli et al., 2021). Transportation planning is a strategic aspect of regional growth and prosperity that is intimately connected to policy formulation and implementation (Huang et al., 2019; May et al., 2017). It is therefore essential that approaches be adopted for monitoring and modeling system performance in order to better inform policies and regulations associated with transport services (Chowdhury & Ceder, 2016; Guzman et al., 2020; Scorcia & Munoz-Raskin, 2019). For example, in the United States, the Clean Air Act and subsequent amendments to it are policy-based regulations aimed at reducing the environmental impacts of emissions (Gumbo & Moyo, 2020). Associated with this has been the continued development of analysis approaches for identifying and assessing alternatives for achieving mandated reductions in emissions. Another example is the urban transportation system, which consists of interrelated models for predicting travel demand (Chakwizira et al., 2019, 2021; Mbara & Pisa, 2018). This is essential for evaluating impacts on facilities and various modes of travel resulting from regional growth. A final example is the interaction between land-use and transportation which has been the subject of modeling efforts. The key component in the emerging methods and the techniques for better understanding transportation processes is geographical information systems (GIS).

The data generated by mobile applications which rely on location sensors have the potential to analyze the day-to-day movement networks of commuters. However, in analyzing this data, setbacks were identified (Li et al., 2021; Žunić et al., 2020). For example, the information measured was subject to noise and uncertainties, hence leading to imprecise results, if these were not excluded in the analysis (Vascik & Hansman, 2018). The analysis of social media data can best be expressed through an insight that has evolved over the years. For instance, Moyo and Musakwa (2016) mention that "scholars between 1995 to 2009 analyzed data as a means to an end, while those between the years 2009 to 2013 analyzing data as both a means to an end and also as the end". Currently, there has been a paradigm shift, with scholars from 2013 to 2016 analyzing data as 'the end' and those post-2014 analyzing data as a service.

Southern African metropolitan cities have actively participated in the investment and development of the Bus Rapid Transit (BRT) Systems such as MyCiTi in Cape Town, Rea Vaya in Johannesburg, and Are Yeng in Tshwane. In South Africa, the introduction of BRT systems was officially sanctioned in the Department of Transport (DoT) Public Transport Strategy and the Public Transport Strategy Action Plans of 2007 (Walters, 2013). The BRT Program was designed to move large numbers of people to all parts of a city quickly and safely, through systems linking different parts of a city into a network to ensure that most residents are no more than 500 m away from stations (Venter et al., 2019).

The system's feature has dedicated bus-only lanes and bus stations that are safe, comfortable, protected from the weather, and friendly to commuters with special needs. In support of the BRT systems, the Gauteng Metropolitan government participated in the investment and development of a high-speed rail network, the ‘Gautrain’, which caters to all of its three metropolitan municipalities. The Gauteng provincial government has also developed the ‘Gaubus’, a significant bus feeder and distributor system which supports the rail service to and from stations. The Gautrain system interfaces with Metrorail commuter services at the Hatfield, Pretoria, Rhodesfield, and Johannesburg Park Stations.

There have nevertheless been rising concerns in terms of the state of public transport systems. One of these has been the lack of well-integrated, reliable, and efficient public transport systems. Specifically, this chapter explores how the Internet of Things can improve public transportation systems in the developing world. The chapter focuses on the state of public transport systems in four South African metropolitan cities: Cape Town, Johannesburg, Tshwane, and Ekurhuleni. Travel time data, trip planning data, payment methods data, GeoWeb 2.0 data, and GTFS data gathered during the study were used to unpack planning and the state of integration of urban public transport systems.

4.2 Indicators of Integration

Table 4.1 depicts the results from the crowdsourced data. The indicators were mostly analyzed using quantitative analysis and semantic analysis of emoticons (Gal-Tzur et al., 2014). Indicators that were not quantitative such as automated and electronic automatic systems, information displays, and mobile applications, were classified as yes if they were available or and no if not—emoticons were then used to describe the level of service.

 represents yes/positive review; while  represents no/negative review.

4.3 Network Analysis

The Gautrain is an advanced rapid rail system, and the Gaubus is a Bus Rapid Transit system that serves as an extension of the Gautrain between Johannesburg, Tshwane, and Ekurhuleni. Gauteng (Fig. 4.1) is South Africa’s economic hub and currently experiences traffic congestion on its major routes, mainly between Pretoria and Johannesburg. The current transport facilities and services between these two cities are mainly road-based. Hence the Gautrain was supposed to ease this traffic congestion to create a smart city based on mixed land uses and development corridors. A total of 46% of commuters used Gautrain in the early morning peak hours (04:50–06:29), 39% of the commuters use the Gautrain mid-morning (06:30–08:30), while 4% of the commuters use the Gautrain in mid-afternoon (08:30–15:30), 7% late afternoon (15:30–18:00), 4% evening peak hour (18:00–21:30). The results reveal

Table 4.1 Indicators on smart mobility and uptake of public transport

Variable	Indicators		Rea Vaya	Gautrain/ Gaubus	MyCiti	A Re Yeng	Harambee	Mini-bus taxi in Bulawayo
Internet of Things	Automated and electronic ticket system		☹	☺	☺	☹	☹	☹
	Availability of route schedules and waiting time		☺	☺	☺	☺	☺	☹
	Payment options	Electronic ticket system	☺	☺	☺	☺	☺	☺
		One ticket system	☹	☺	☹	☹	☹	☹
		Bank card	☺	☺	☺	☺	☺	☹
		App payment	☹	☺	☺	☹	☹	☹
	Notification services		☹	☺	☺	☹	☹	☹
	Presence on Web 2.0 platforms	Twitter	☺	☺	☺	☺	☺	☹
		Facebook	☺	☺	☺	☺	☺	☺
		Own app	☹	☺	☺	☹	☹	☹
Public transport uptake	Network density		☹	☹	☹	☹	☹	☺
	Demand for public transit		☺	☺	☺	☹	☹	☺
	Average number of commuters		45,000 per day	580,000 per month	66,000 per day	12,000 per day	8000 per day	600,000 per month
	Level of integration		☹	☺	☹	☹	☹	☺
	Robustness of network		☹	☹	☹	☹	☹	☺
Industry 4.0	Route network	Open source	☹	☺	☹	☹	☹	☹
	data availability	Available on request	☺	☺	☺	☺	☺	☺
	Are the public transportation providers adept and ready for Industry 4.0?		☹	☺	☺	☹	☹	☹
	How far are the public transportation providers in digitalisation of operations		0.30	0.60	0.50	0.30	0.20	0.15

Source Arnold et al. (2017), Fombad (2015), Musakwa and Gumbo (2017), Van Der Westhuizen (2007)

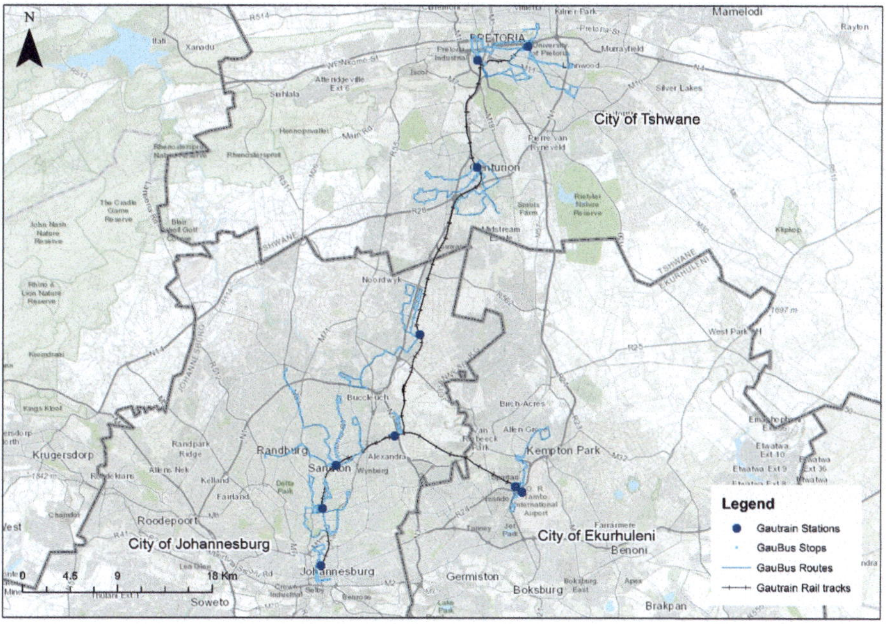

Fig. 4.1 Gautrain and Gaubus route map in the cities of Johannesburg, Tshwane, and Ekurhuleni

that most people utilize the Gautrain for home to work trips. This is also reflected by the train stations located in economic nodes of the four cities, as reflected through the Spatial Development Frameworks.

The Rea Vaya operates in Region A to F in the Johannesburg Metropolitan City in different phases and has systematic hierarchical routes connecting micro-city centers in the city. It has completed the construction of Phase 1A and 1B and is currently developing Phase 1C. Rea Vaya's Phase 1A has a trunk route operating between Ellis Park in Doornfontein and Thokoza Park in Soweto, linking with several feeder routes in Soweto (Rea Vaya, 2015). Feeder buses run from Protea Glen to Thokoza Park and Eldorado Park to Lakeview The route covers 325 km of special lanes and intersections, while feeder and complementary buses carry passengers to the trunk route stations. The inner city circular route travels around the CBD, from Hillbrow and Braamfontein to Ellis Park in the east and Chancellor House on the western edge of the city (see Fig. 4.2).

Phase 1B has routes operating through Cresta, Windsor West, Parktown, Yeoville; in addition, routes to and from the University of Johannesburg Soweto are being added. The route starts in Noordgesig in Soweto and travels through Pennyville, New Canada, Highgate, Auckland Park, and Braamfontein to Parktown, Metro Center, and Rissik Street in the CBD. The route has made it possible for commuters to reach critical public healthcare centers such as the Rahima Moosa, Helen Joseph, and Charlotte Maxeke hospitals. Educational institutions such as the University of Johannesburg, University of the Witwatersrand, Milpark College, Parktown Boys'

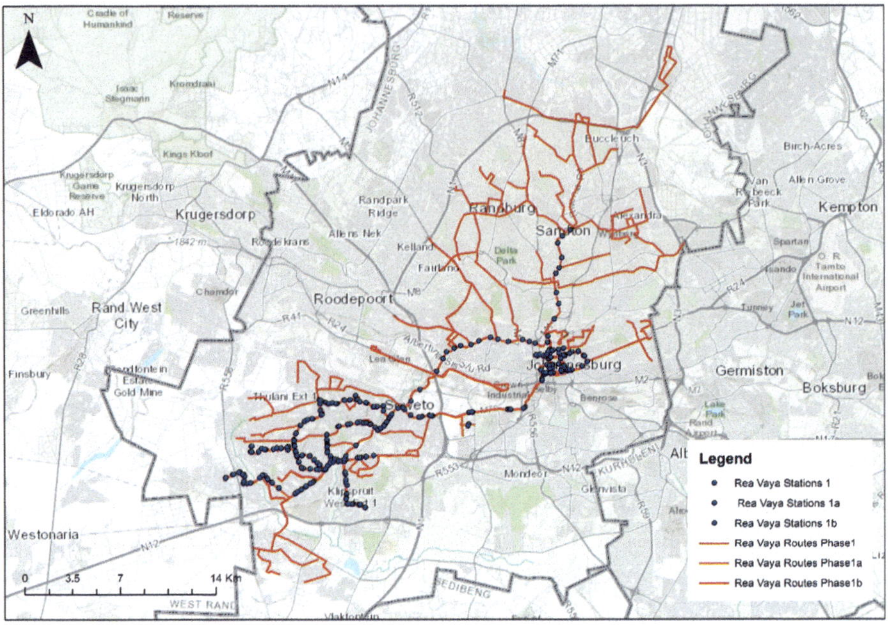

Fig. 4.2 Rea Vaya route map in the city of Johannesburg

High School, and Barnato Park High School are also accessible. Feeders run to and from Leaglen, Stormhill, Florida, Cresta, Yeoville, and Parktown. There are also additional feeders in Soweto from Pimville and Mapetla. These routes now link to the Metro Center Rea Vaya loop, which travels to the inner city through Braamfontein.

Generally, transportation nodes with high centrality allow the flow of commuters between every pair of bus stops over the shortest paths between them. For the Harambe network (see Fig. 4.3), nodes along the Bus Route are currently located within close proximity to economic nodes, leading to a positive public transportation uptake. This could be due to the service quality of the Bus Rapid Transit system, as most commuters utilize the bus to travel in between townships and the Metrorail for longer trips.

The new City of Tshwane A Re Yeng BRT system is transports commuters across the city efficiently by using specially designated lanes with enclosed bus stations. The hub is located at the corner of Nana Sita and Paul Kruger streets in the city center. The network features two circular routes traversing the Pretoria CBD and various trunk routes linking to Pretoria's university campuses, with further routes currently under construction. The significance of this system is to assist in transforming spatial apartheid planning. Further, to use the bus, commuters need to register for a connector card (for a once-off fee of R25) at any A Re Yeng bus station (see Fig. 4.4). They then proceed to the turnstile, and after swiping the card, receive a paper ticket that is later shown to conductors during the journey.

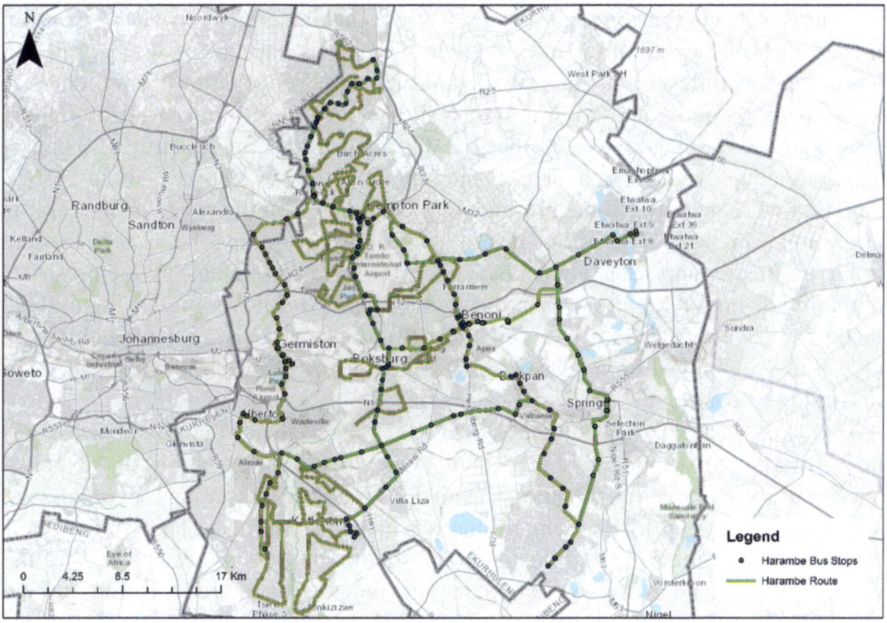

Fig. 4.3 Harambe route map in the city of Ekurhuleni

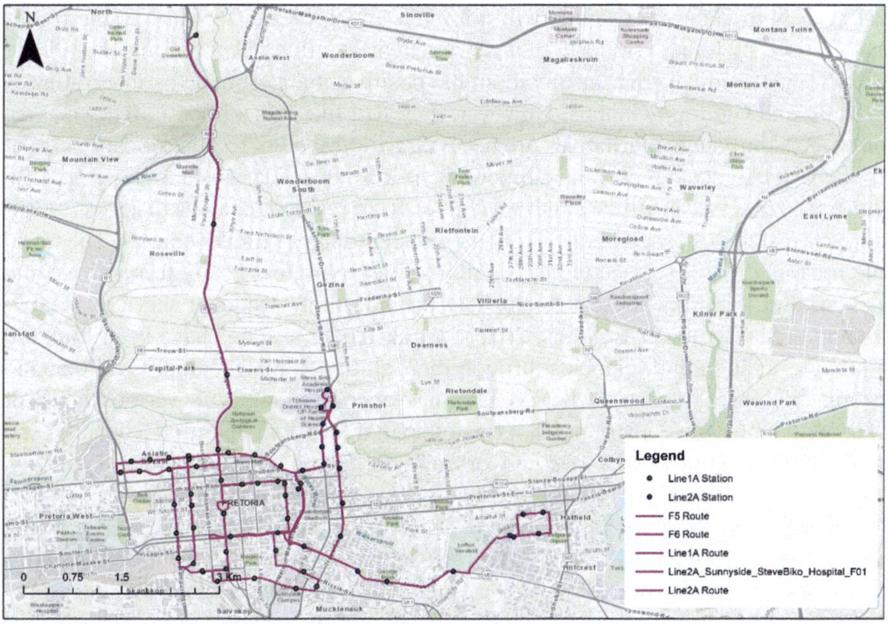

Fig. 4.4 Are Yeng route map in the city of Tshwane

Generally, A Re Yeng officials are optimistic about the services provided. Approximately 50% of the responses indicate that commuters have access to information about time schedules. Maps are provided at the A Re Yeng stations for commuters to plan trips properly, and there is information on the A Re Yeng website online. Consequently, officials indicated that bus fare collections are electronic, using e-smart cards deployed through a tag in and tag out at the bus stations. About 25% of responses were negative, though, indicating challenges on the road, such as traffic congestion which resulted in bus delays. On a few occasions, the system loading money on the e-smart card goes offline due to system maintenance which restricts commuters from loading money on the e-smart cards. Further, officials highlighted that the routes serviced by the buses are mostly the busier ones. Hence, potential route extensions are planned to increase the A Re Yeng bus services to other locations in the City of Tshwane.

About 70% of A Re Yeng commuters gave negative feedback that payment systems and information distribution are inadequate; bus arrivals at the stations do not correspond to the bus timetable schedules provided. This can be frustrating for commuters, as they cannot use the A Re Yeng at that particular moment and have to use alternatives to get to their desired destinations. Spatially, 40% of feedback from the commuters indicated that A Re Yeng transports them to the desired locations around the CBD. These were mostly from the University of South Africa (Unisa) students and residents residing and working in Tshwane. About 40% of the feedback from commuters indicated that A Re Yeng does not go to the desired locations and integrating it with Gautrain/Gaubus will not make much of a difference, as the routes that are used by both systems do not service their locations. Approximately 10% of A Re Yeng commuters are unsure about the positive changes made by integrating A Re Yeng and Gautrain/Gaubus. However, they feel like it is a good initiative, as on some occasions such alternatives are necessary for traveling to other locations.

The My Citi system currently functions as a trunk-feeder system (see Fig. 4.5). Trunk-feeder services utilize smaller vehicles from residential areas to access terminals or transfer stations, where customers transfer to larger trunk vehicles. The rationale for this type of service is that smaller vehicles are less costly to purchase and operate, and therefore these vehicles can provide more frequent services in low-demand areas (Bruun & Behrens, 2016). Trunk-feeder systems have been found to improve operational efficiencies through their ability to match supply and demand (Bruun & Behrens, 2016). All MyCiTi lines operate along with various types of roads which provide constraints in terms of spatial location, direction, capacity, and operating speed. In EMME, road networks are built from links and nodes. Typically, roads consist of two links to indicate two opposing directions of travel.

According to the City of Bulawayo, the city's busiest roads with large traffic volumes are Plumtree Road, Victoria Falls Road, Khami Road, Burnside and Matopos Roads, and Luveve Road (Ndlovu & Newman, 2021). Three recognized public transport companies operate vans (combis) as a vehicle of transportation. The average occupancy of each van during peak hours is 18 passengers. According to the city council, the plan was that these three companies would eventually merge into and shift their mode of operational vehicles from vans to 25-seater minibusses or more

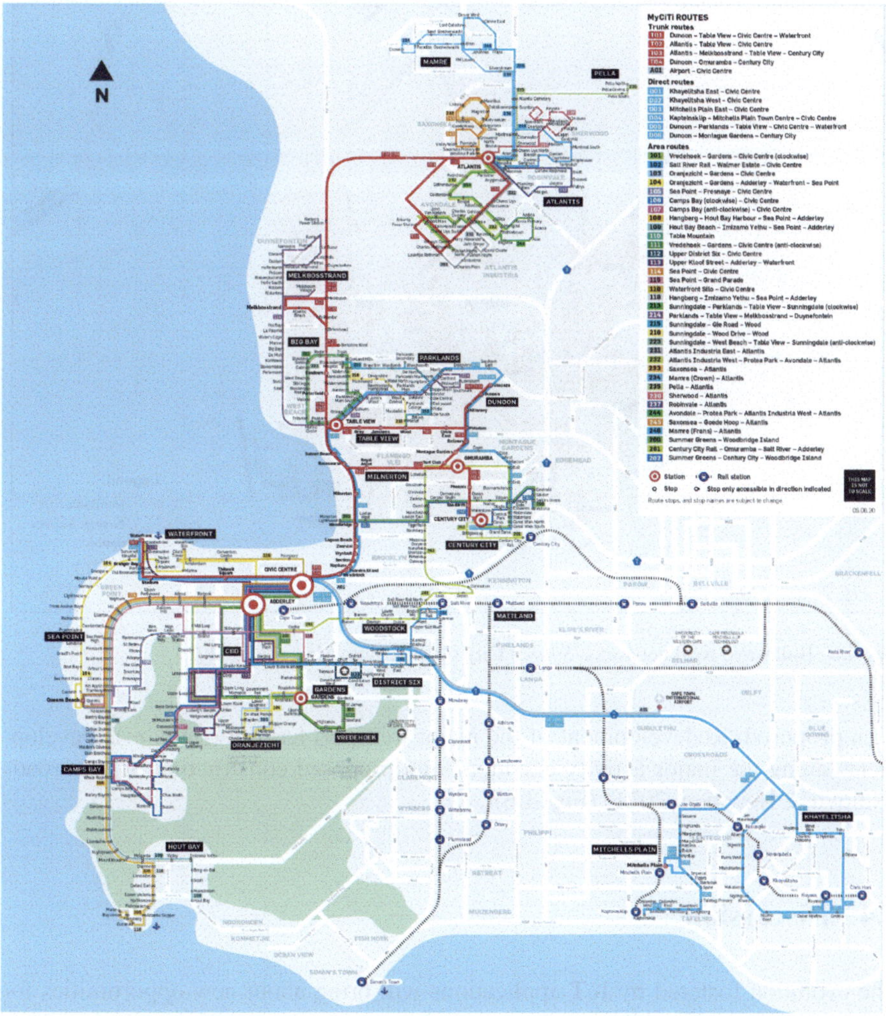

Fig. 4.5 Myciti route map in the city of Cape Town. *Source* https://www.myciti.org.za/en/routes-stops/route-map-downloads/

ideally, introduce standard buses. Due to their small sizes, the vans do random stops, and they pull off onto the road shoulders to drop off and pick up passengers. This kind of journey, demand-based operation damages road shoulders, compromises the safety of passengers, and makes travel time much longer than transit on fixed routes, especially if given priority. Moreover, despite having three registered public transport companies, many pirate operators are still operating from small private cars. These small pirate operators have an exploitive reputation and are potentially dangerous, so they are a significant hazard to the safety of public commuters. To improve mobility in the city, the following major development proposals for the City of Bulawayo have

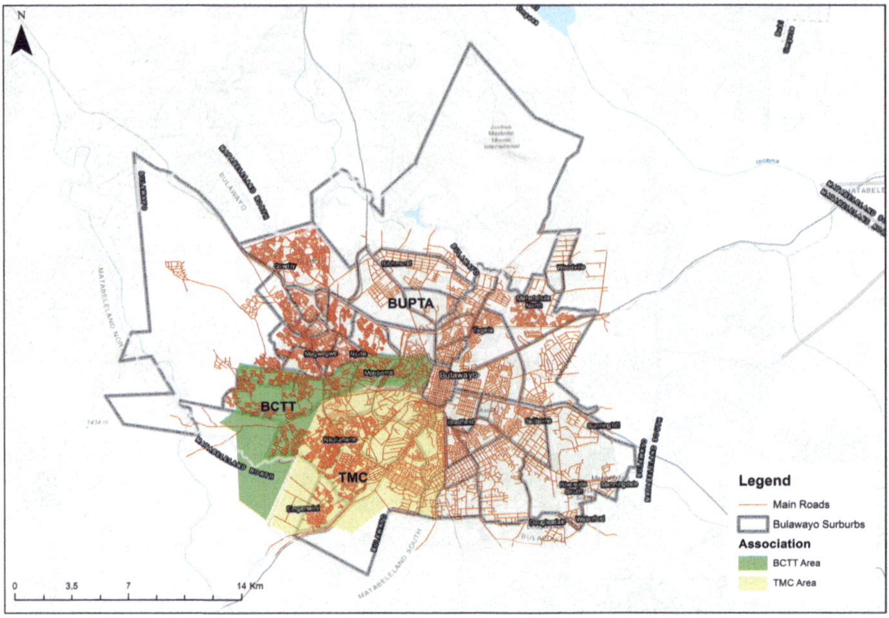

Fig. 4.6 Bulawayo road networks. *Source* City of Bulawayo Master Plan 2019

been proposed: Redevelopment of the major transport termini (Egodini), Developments along the major activity corridors, and proposed construction of new roads (Ndlovu & Newman, 2021) (Fig. 4.6).

4.4 Discussion

The evolution fostered by IoT applications will bring about new opportunities for nearly all aspects of public transport, including opportunities of interest to strategic transport planners and opportunities of interest to operators. These two types of opportunities will enable new innovative mobility models and positively impact the metropolitan cities' contexts. A significant opportunity that is currently untapped is the increasing number of transport commuters. Following the analysis, it became apparent that to gain business from such commuters, a critical aspect for improving transportation is to properly adjust and balance public transport services, improving the quality of services for all segments of the population, and increasing the attractiveness of public transport. In light of this, IoT technologies can be leveraged to provide the necessary data to improve transport usage monitoring and understand usage patterns, eventually allowing for dynamically adjusting and balancing transport services.

4.5 Implications, Reflections, and Generalizability of Findings

The chapter's contribution to existing knowledge about traffic management is twofold. First, the crowdsourced data is presented to estimate commuters' experiences and observations to assist traffic management policy. Crowdsourcing has been used in several case studies to measure, predict or evaluate citizen experiences and observations from developed and developing countries. Secondly, the route network of the public transportation providers is assessed. To the researcher's knowledge, all the innovative public transportation systems are still expanding their networks, but how the network is assessed has the potential to inform future expansion plans. Lastly, this chapter supports previous chapters' efforts to evaluate the status quo of existing public transportation systems. In the integration of various modes, the components of the cost are not in general binary values. As such the research used several public transportation systems to outline the implications and robustness of the networks. Lessons learned reveal that innovative public transportation systems can integrate with traditional modes, which can be extended to other developing cities.

4.6 Conclusion

This chapter has explored how the Internet of Things can improve public transportation systems in the developing world. It has focused on the state of public transport systems in four South African metropolitan cities: Cape Town, Johannesburg, Tshwane, and Ekurhuleni. Travel time data trip planning data, payment methods data, GeoWeb 2.0 data, and GTFS data gathered during the study were used to unpack planning and the state of integration of urban public transport systems. The evolution fostered by IoT application will bring about new opportunities for nearly all aspects of public transport, including opportunities of interest to strategic transport planners and to operators. The two types of opportunities enable new innovative mobility models and positively impact the metropolitan cities' contexts. A significant opportunity that is currently untapped is the increasing number of transport commuters. Following the analysis, it has become apparent that a critical aspect for improving transportation is to properly balance public transport services, improving the quality of services for all segments of the population and increasing the attractiveness of public transport in order to gain new commuters.

References

Arnold, K., Le Roux, A., & Hattingh, M. (2017). Impact of Gautrain stations on property prices and sales activity in the City of Johannesburg between 2006 and 2015. *South African Journal of Geomatics, 6*, 184–195.

Bruun, E., & Behrens, R. (2016). Paratransit in Sub-Saharan African cities: Improving and integrating informal services. In *Paratransit: Shaping the flexible transport future*. Emerald Group Publishing Limited.

Chai, C., Fan, J., Li, G., Wang, J., & Zheng, Y. (2019). Crowdsourcing database systems: Overview and challenges. In *2019 IEEE 35th International Conference on Data Engineering (ICDE)* (pp. 2052–2055). IEEE.

Chakwizira, J., Bikam, P., & Adeboyejo, T. A. (2019). Transport realities and challenges for low income peripheral located settlements in Gauteng Province: Are we witnessing the genesis of a new transport order or consolidation of the old transport order? In: *International Conference on Applied Human Factors and Ergonomics* (pp. 618–630). Springer.

Chakwizira, J., Bikam, P., & Adeboyejo, T. A. (2021). Access and constraints to commuting for persons with disabilities in Gauteng Province, South Africa. In *Urban inclusivity in Southern Africa*. Springer.

Chowdhury, S., & Ceder, A. A. (2016). Users' willingness to ride an integrated public-transport service: A literature review. *Transport Policy, 48*, 183–195.

Fombad, M. C. (2015). Governance in public–private partnerships in South Africa: Some lessons from the Gautrain. *Journal of Southern African Studies, 41*, 1199–1217.

Gal-Tzur, A., Grant-Muller, S. M., Kuflik, T., Minkov, E., Nocera, S., & Shoor, I. (2014). The potential of social media in delivering transport policy goals. *Transport Policy, 32*, 115–123.

Gumbo, T., & Moyo, T. (2020). Exploring the interoperability of public transport systems for sustainable mobility in developing cities: Lessons from Johannesburg Metropolitan City, South Africa. *Sustainability, 12*, 5875.

Guzman, L. A., Arellana, J., & Alvarez, V. (2020). Confronting congestion in urban areas: Developing sustainable mobility plans for public and private organizations in Bogotá. *Transportation Research Part a: Policy and Practice, 134*, 321–335.

Huang, X., Cao, X. J., Yin, J., & Cao, X. (2019). Can metro transit reduce driving? Evidence from Xi'an, China. *Transport Policy, 81*, 350–359.

Kemajou, A., Jaligot, R., Bosch, M., & Chenal, J. (2019). Assessing motorcycle taxi activity in Cameroon using GPS devices. *Journal of Transport Geography, 79*, 102472.

Li, V. O., Lam, J. C., Han, Y., & Chow, K. (2021). A big data and artificial intelligence framework for smart and personalized air pollution monitoring and health management in Hong Kong. *Environmental Science & Policy, 124*, 441–450.

Marzano, G., Lizut, J., & Siguencia, L. O. (2019). Crowdsourcing solutions for supporting urban mobility. *Procedia Computer Science, 149*, 542–547.

May, A., Boehler-Baedeker, S., Delgado, L., Durlin, T., Enache, M., & van der Pas, J.-W. (2017). Appropriate national policy frameworks for sustainable urban mobility plans. *European Transport Research Review, 9*, 7.

Mbara, T., & Pisa, N. (2018). An analysis of impediments to deliver sustainable transport in cities of the developing countries: The case of Harare, Zimbabwe.

Moyo, T., & Musakwa, W. (2016). Using crowdsourced data (Twitter & Facebook) to delineate the origin and destination of commuters of the Gautrain public transit system in South Africa.

Musakwa, W., & Gumbo, T. (2017). *Impact of urban policy on public transportation in Gauteng, South Africa: Smart or dumb city systems is the question*. Springer.

Musakwa, W., & Selala, K. M. (2016). Mapping cycling patterns and trends using Strava Metro data in the city of Johannesburg, South Africa. *Data in Brief, 9*, 898–905.

Ndlovu, V., & Newman, P. (2021). How would the trackless tram system and Public-Private Partnership (PPP) apply to Bulawayo? *Current Urban Studies, 9*(1), 17–30.

Scorcia, H., & Munoz-Raskin, R. (2019). Why South African cities are different? Comparing Johannesburg's Rea Vaya bus rapid transit system with its Latin American siblings. *Case Studies on Transport Policy, 7*, 395–403.

Sun, Y., Du, Y., Wang, Y., & Zhuang, L. (2017). Examining associations of environmental characteristics with recreational cycling behaviour by street-level Strava data. *International Journal of Environmental Research and Public Health, 14*, 644.

Van Der Westhuizen, J. (2007). Glitz, glamour and the Gautrain: Mega-projects as political symbols. *Politikon, 34*, 333–351.
Vascik, P. D., & Hansman, R. J. (2018). Scaling constraints for urban air mobility operations: Air traffic control, ground infrastructure, and noise. In *2018 Aviation Technology, Integration, and Operations Conference* (p. 3849).
Vecchio, G., & Tricarico, L. (2019). "May the Force move you": Roles and actors of information sharing devices in urban mobility. *Cities, 88*, 261–268.
Venter, M., Steynberg, L., Grundling, J. P., & Feng, X. (2019). Perceived contribution of township enterprises on local economic development in Mabopane Township, South Africa. *Journal of Public Administration, 54*, 888–907.
Vermicelli, S., Cricelli, L., & Grimaldi, M. (2021). How can crowdsourcing help tackle the COVID-19 pandemic? An explorative overview of innovative collaborative practices. *R&D Management, 51*, 183–194.
Walters, J. (2013). Overview of public transport policy developments in South Africa. *Research in Transportation Economics, 39*, 34–45.
Žunić, E., Delalić, S., & Đonko, D. (2020). Adaptive multi-phase approach for solving the realistic vehicle routing problems in logistics with innovative comparison method for evaluation based on real GPS data. *Transportation Letters*, 1–14.

Chapter 5
Integration of Urban Mobility Systems in the Gauteng City Region

Abstract There has been growth in mobility research focused on improving connectivity between various modes of public transport, particularly in the developing world. Of importance in the Fourth Industrial era is the emerging research on shared mobility services, commonly referred to as multi-modal transportation. Generally, the problem to be solved is to map how commuters traverse from bus stop A to bus stop B. This is solved in a South African context, where innovative public transportation has been developed but without clear integration with traditional public transportation modes. Therefore, there is a need for more robust networks, which provide several alternative routes between various interchange zones, however, creating new routes could have significant economic and ecological costs. This chapter, therefore, uses efficient, societal, ecological, and economic constraints to measure connectivity levels at interchange zones. In the analysis, we use indicators to determine variations of connectivity levels across the Gauteng City Region. The data reveal hot spots of high interchange zones at major mobility hubs. In conclusion, improving connectivity between various modes of public transportation will lead to increased ridership of public transportation.

Keywords Connectivity · Public transportation · Multi-modal · Interchange zones · Networks

5.1 Introduction

Modern cities development and public transportation infrastructure planning have become entwined with technology in the Fourth Industrial era (4IR). This has led to a symbiotic relationship, whereby development planning relies on technology, and technology relies on continuous development (Ge et al., 2018; Hill et al., 2017). With the aid of big data and the internet of things, cities commuters are alerted about possible congestion, hazards on roads, and possible alternative routes (Hussein et al., 2018; Kumar et al., 2019). Such technology has led to comprehensive solutions which increase mobility and integrate public transportation facilities and services. As a consequence, many such cities are moving toward integrating services with technology through real-time traffic and accident monitoring and physical integration of

T. Gumbo et al., *Urban Public Transport Systems Innovation in the Fourth Industrial Revolution Era*, https://doi.org/10.1007/978-3-030-98717-6_5

public transport. For example, in Europe, Copenhagen, Denmark has implemented a host of solutions to make their city more friendly and convenient for cyclists (Gössling & Choi, 2015; Larsen, 2017). In China, Hong Kong uses real-time monitoring from government departments to help travelers navigate their city efficiently (Li et al., 2021; Wan et al., 2018). The government of Chennai, India has also successfully converted motorists to cyclists to cut down on private transportation (Krishankumar et al., 2021; Ponkshe & Pricing, 2020).

This chapter serves as an extension of the previous ones by assessing the physical integration of urban mobility systems. It also adds to the existing body of knowledge through studies that classified integration into the following types: Fare or functional integration, Modal integration, Cross city modes or network integration, Timetable and information integration, Institutional integration, and Policy integration (Aziz & Mohamad, 2020; Rahman, 2013; Takahashi, 2017; Toro et al., 2019). Generally, the problem to be solved is to map how commuters' traverse from bus stop A to bus stop B. The problem is solved in a South African context, where innovative public transportation has been developed but without clear integration with traditional public transportation modes (Chakwizira, 2020; Gumbo & Moyo, 2020; Gumbo et al., 2021; Loubser et al., 2021). Therefore, there is a need for more robust networks, which provide several alternative routes between various interchange zones, as creating new routes could have significant economic and ecological costs. This chapter uses efficient, societal, ecological, and economic constraints to measure connectivity levels at interchange zones.

5.2 Modeling the Integration of Urban Mobility Systems

Using case studies, the chapter then considers augmentation of the following public transportation modes Gautrain, Gaubus, Rea Vaya, Are Yeng, Harambe, and mini-bus taxi in the Gauteng City Region which comprised of the metropolitan municipalities namely City of Johannesburg, City of Tshwane, City of Ekurhuleni (see Fig. 5.1).

The chapter is concerned with assessing the physical integration of public transportation networks, namely innovative public systems and traditional public systems (Rahman, 2013). Inspired by the topologies of innovative public transportation networks encountered in developing countries, the following flow chart (see Fig. 5.2) is an extension of a model developed (Moyo et al., 2021).

The innovative public transportation network represented by Gautrain, Gaubus, Rea Vaya, Are Yeng, Harambe, and mini-bus taxi in the Gauteng City Region (City of Johannesburg; City of Tshwane; City of Ekurhuleni) is modeled with a weighted path graph $Y(V_y, E_y, W_y)$; where V_y, E_y, and W_y represent the vertex (nodes or stops) set, the edges set (link connecting nodes), and the weights set, respectively (see Fig. 5.3).

Given that mini-bus taxis are a common mobility mode in developing countries, the chapter proposes their utilization in improving commuting within the Gauteng City Region (Govender, 2016). The weighted path graph for the mini-bus taxi is

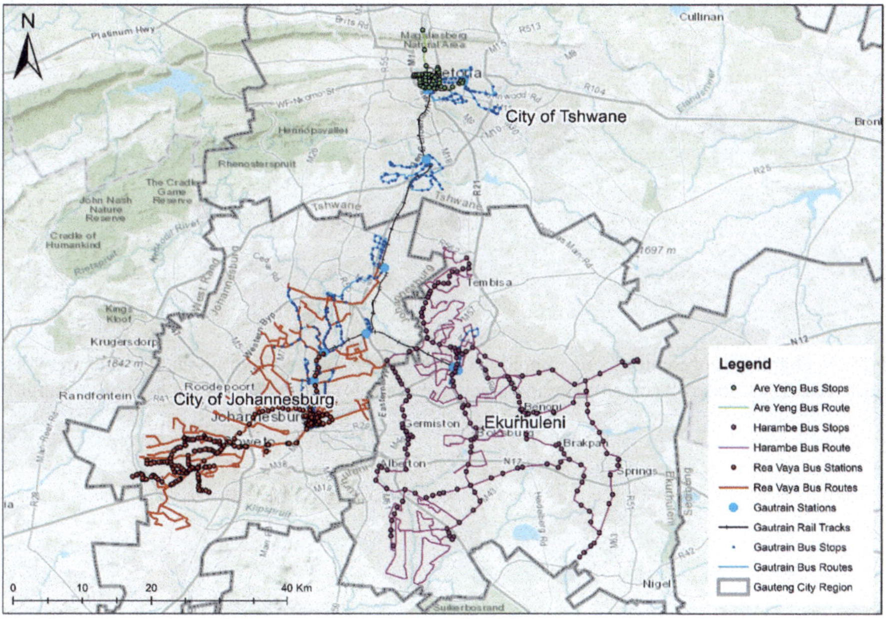

Fig. 5.1 Gauteng city region

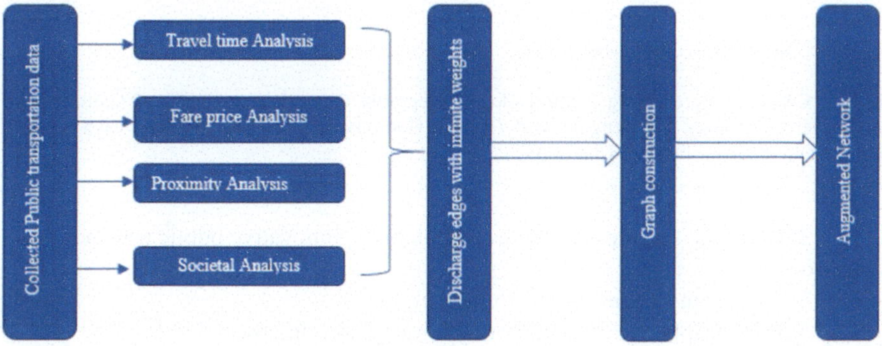

Fig. 5.2 Flowchart. Adapted from Moyo et al. (2021)

defined as $Z(V_z, E_z, W_z)$; where V_z, E_z, and W_z represent the vertex (nodes or stops) set, the edges set (link connecting nodes), and the weights set, respectively (see Fig. 5.4).

Therefore, to solve the problem in Matlab, the approach is divided into five steps

Step 1: Given public transportation networks, perform a cost analysis to identify parameters.

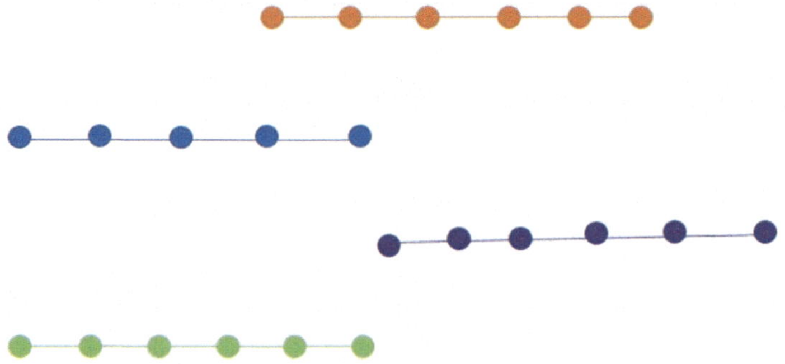

Fig. 5.3 Graph Y representing a disjointed network comprising of Gautrain and Gaubus network (blue); Are Yeng network (green); Harambe network (purple) and the Rea Vaya network (orange)

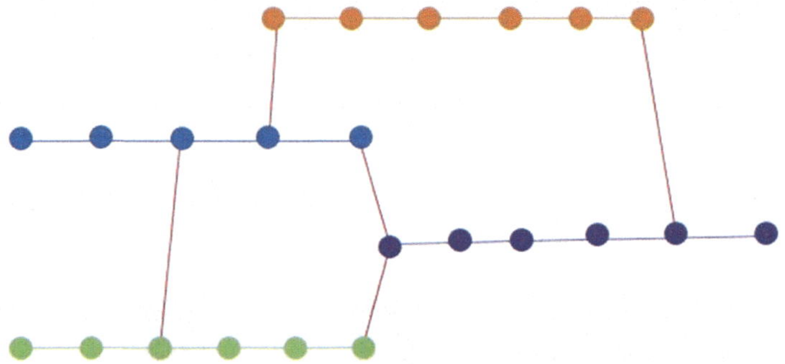

Fig. 5.4 Graph Z representing a multi-modal integrated network comprising of Gautrain and Gaubus network (blue); Are Yeng network (green); Harambe network (purple); Rea Vaya network (orange) and the Mini-bus taxi (maroon)

Step 2: Build the set of possible new edges to the innovative public transportation network.
Step 3: Discarding edges with infinite weights.
Step 4: Find graph $Y \sqcup Z$, with corresponding weight cost.
Step 5: Augment the innovative public transportation graph with mini-bus edges.

5.3 Modeling of Augmented Network

The public transportation networks are represented as a disjointed union of graphs, whereby each public transportation network forms its own disjointed root tree. The problem to be solved is to merge these networks using mini-bus taxis operations. The new integrated network will therefore be represented by $G(V, E, W) = Y \sqcup Z$, Y defined as the graph for the innovative public transportation network and Z defined as the graph for the mini-bus taxis.

Currently, due to the structure of Y, the innovative public transportation networks operate in isolation, and any link failure in Y will further create isolated components. The chapter proposes the augmentation of the graph Y by adding new edges to make the graph more robust. The new edges to be added will include the mini-bus taxis network which is already widely utilized, however whose operations are not fully regulated (Govender, 2016; Sebola, 2014). Our problem is to merge graphs Z and Y; usually, the cost is related to travel time metrics but, herein a composite cost is adopted which includes travel time, proximity, fare per mode, and societal constraints. This cost is the sum of weights defined hereafter. Considering a transportation network represented with a weighted graph. The edge weight between vertices i and j using mode m is given by:

$$w_{ijm} = t_{ijm} p_{ijm} s_{ijm} f_{ijm}$$
$$\text{Equation 1 : Weighted graph} \tag{5.1}$$

where

t_{ijm} represents the travel time between i and j when using mode m.
p_{ijm} represents the proximity/accessibility cost between i and j using mode m.

For an innovative public transportation mode m, if there is a direct link between i and j, then $a_{ijm} = 1$, or else $a_{ijm} = \infty$. For the mini-bus taxi, if the distance between i and j is lower than a given threshold D, and there is a mini-bus taxis route then $p_{ijm} \geq 1$, or else $a_{ijm} = \infty$. This cost can also consider walkability of the link between i and j in order to consider the additional distance commuters will walk to and from the various mode stops.

The quantity s_{ijm} represents the security (safety) and preference parameter cost between i and j using mode m. It allows for the consideration of the crime level in the traversed zone. We assume the innovative public transportation stations are safe due to the additional security at these locations (such as covered bus stops and security personal). Therefore, $s_{ijm} = 1$ for innovative public transportation modes. For mini-bus taxi, $s_{ijm} \geq 1$. Here, various perceptions of safety can be included. One can for instance set the cost to ∞ if the link is not safe at all from a statistical point of view or from commuters' perceptions.

The quantity f_{ijm} represents a fare per mode which is equal to 1 for innovative public transportation modes as they have multi fare payment methods. For mini-bus taxi, $p_{ijm} = \infty$, as these require cash, if commuters are willing to pay the fare using cash between i and j then $f_{ijm} \geq 1$.

Generally, nodes with high closeness centrality scores are within the path with the shortest distance to all other nodes in the network. Using the reclassification tool in Arc GIS to normalize the results, similarities and differences between the two graphs can be seen. Using the Jenks natural breaks algorithm in the reclassification tool, class breaks were created between a range of 1–10. This allowed for grouping of similar values together while also maximizing the differences between class ranges. Nodes with a score of 9 or 10 in this work are referred to as 'central nodes'. After introducing new edges, the nodes of innovative public transportation nodes connectivity was improved (see Fig. 5.5).

Figure 5.6 represents the sets of possible new edges to be added to the innovative public transportation network. After introducing the composite cost, edges with an

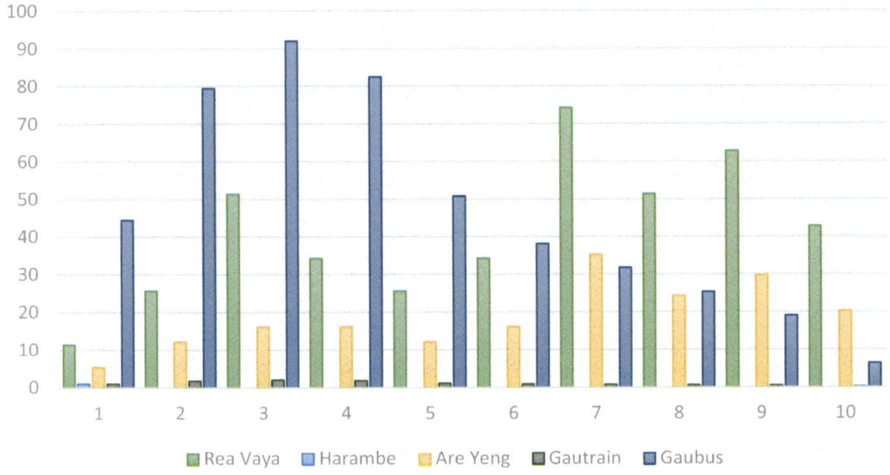

Fig. 5.5 Centrality score

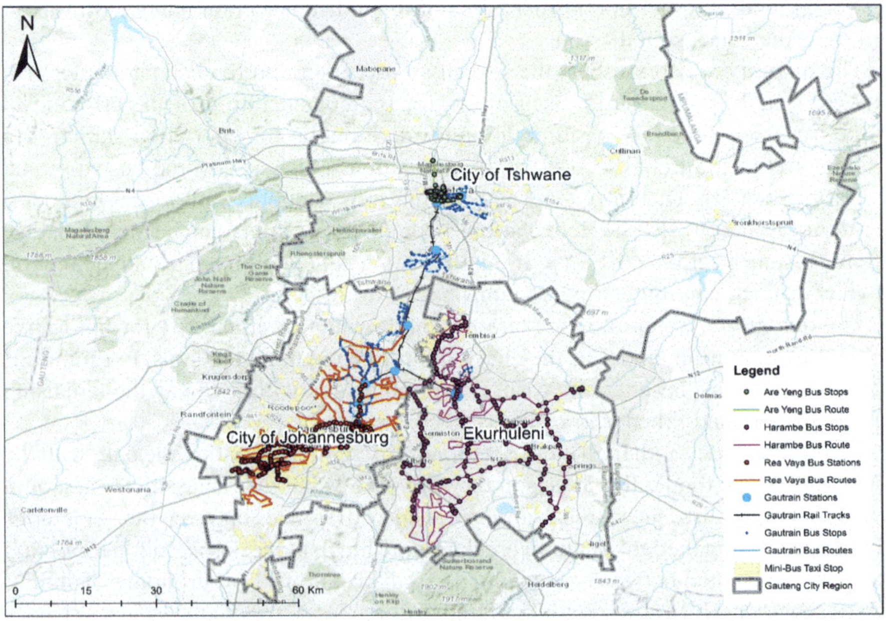

Fig. 5.6 Augmented network

infinite cost are discarded. It must be stressed that urban mobility is essential for the maintenance and wellbeing of society and economy. As such, improving mobility with the Gauteng City Region has potential benefits which will spread far beyond the transportation sectors, since it is essential for economic development and improving the quality of life of people.

5.4 Implications, Reflections, and Generalizability of Findings

The proposed mathematical solution of network augmentation given reveals how to physically integrate innovative public transportation with the traditional modes (mini-bus taxi). The approach was assessed within a developing country, where new innovative public transportation and traditional modes are not fully integrated within the public transport network. The framework proposed in this chapter sought to unpack how an innovative public transportation network can be made more robust by determining the stations and routes whereby mini-bus taxis routes can be added to improve connectivity.

In the integration of various modes, the components of the cost are not in general binary values. As such, the chapter used several scenarios to test the implications and robustness of the augmented network. The results reveal that mini-bus taxis can be used in developing countries to connecting commuters to several points of interest in the city given their existing robust network. Furthermore, lessons learned from assessing the impact of merging innovative public transportation and traditional modes can be extended to other developing cities. Given the existing robust network of mini-bus taxis, there is a need to formalize their operations through merging their network and operations with emerging modes. Also given this potential, city planners and public transportation providers need to develop and streamline plans on land-use and traffic management to reduce the negative impacts of public transportation, while also improving commuting trips.

5.5 Conclusion

The assessment of the potential for merging innovative and traditional public transportation networks can be used to inform plans to ensure public transportation is highly flexible and demand-responsive while also ensuring its sustainability. For innovative public transportation, services such as in the Gauteng City Region to reduce the negative impact of having not being connected to other systems, there is a need to develop a composite cost that shall inform how they can be connected to other systems. Lastly to augment a disjointed network the chapter suggests using existing traditional modes as a viable option. This is done through building on previous studies to assist developing countries, as the future of public transportation in urban cities lies in the integration of new mobility systems and traditional mainstream public transportation modes.

References

Aziz, A., & Mohamad, J. (2020). Public transport planning: Local bus service integration and improvements in Penang, Malaysia. *Planning Malaysia, 18.*

Chakwizira, J. (2020). The intersection of spatial fragmentation and smart transport planning in Gauteng Province, South Africa: constraints and opportunities. In *International Conference on Applied Human Factors and Ergonomics* (pp. 367–373). Springer.

Ge, M., Bangui, H., & Buhnova, B. (2018). Big data for internet of things: A survey. *Future Generation Computer Systems, 87*, 601–614.

Gössling, S., & Choi, A. S. (2015). Transport transitions in Copenhagen: Comparing the cost of cars and bicycles. *Ecological Economics, 113*, 106–113.

Govender, K. K. (2016). Exploring public transport service quality: The case of mini-bus taxi service in South Africa. *Eurasian Business Review, 6*, 101–116.

Gumbo, T., Ingwani, E., Mufungizi, A. A., & Dumba, S. (2021). Planning for climate and cyclone resilient transport infrastructures in Southern Africa: lessons from Zimbabwe. In *Cyclones in Southern Africa.* Springer.

Gumbo, T., & Moyo, T. (2020). Exploring the interoperability of public transport systems for sustainable mobility in developing cities: Lessons from Johannesburg Metropolitan City, South Africa. *Sustainability, 12*, 5875.

Hill, N., Gibson, G., Guidorzi, E., Amaral, S., Parlikad, A., & Jin, Y. (2017). Scoping study into deriving transport benefits from big data and the internet of things in smart cities.

Hussein, W. N., Kamarudin, L., Hussain, H. N., Zakaria, A., Ahmed, R. B., & Zahri, N. (2018). The prospect of internet of things and big data analytics in transportation system. In *Journal of Physics: Conference Series* (p. 012013). IOP Publishing.

Krishankumar, R., Pamucar, D., Deveci, M., & Ravichandran, K. S. (2021). Prioritization of zero-carbon measures for sustainable urban mobility using integrated double hierarchy decision framework and EDAS approach. *Science of the Total Environment, 797*, 149068.

Kumar, S., Tiwari, P., & Zymbler, M. (2019). Internet of Things is a revolutionary approach for future technology enhancement: A review. *Journal of Big Data, 6*, 1–21.

Larsen, J. (2017). The making of a pro-cycling city: Social practices and bicycle mobilities. *Environment and Planning A, 49*, 876–892.

Li, V. O., Lam, J. C., Han, Y., & Chow, K. (2021). A big data and artificial intelligence framework for smart and personalized air pollution monitoring and health management in Hong Kong. *Environmental Science & Policy, 124*, 441–450.

Loubser, J., Marnewick, A. L., & Joseph, N. (2021). Framework for the potential userbase of mobility as a service. *Research in Transportation Business & Management, 39*, 100583.

Moyo, T., Kibangou, A. Y., & Musakwa, W. (2021). Societal context-dependent multi-modal transportation network augmentation in Johannesburg, South Africa. *Plos one, 16*, e0249014.

Ponkshe, A., & Pricing, C. (2020). *Policymaking towards green mobility in India.* Occasional Papers.

Rahman, M.S.-U. (2013). *Integrating BRT with rickshaws in developing cities: A case study on Dhaka City.* University of Leeds.

Sebola, M. (2014). Recapitalizing mini-bus taxis for effective public transportation in South Africa: The urban rural transport connection problem. *Urban Transport XX, 138*, 125.

Takahashi, T. (2017). Economic analysis of tariff integration in public transport. *Research in Transportation Economics, 66*, 26–35.

Toro, E. E. M., van der Krogt, A., & Flores, R. S. (2019). Mobility and integration of public transport systems in Latin America. In *Proceedings of the 2019 2nd International Conference on Machine Learning and Machine Intelligence* (pp. 58–62).

Wan, W. H., Tsang, Y. T., Zhu, H., Koo, C. H., Liu, Y., & Lee, C. C. T. (2018). A real-time drivers' status monitoring scheme with safety analysis. In *IECON 2018–44th Annual Conference of the IEEE Industrial Electronics Society* (pp. 5137–5140). IEEE.

Part IV
Functional and Systematic Integration of Public Transport Systems

Chapter 6
Functional Integration of Public Transport Systems in Tshwane, Johannesburg, Harare, and Bulawayo Cities

Abstract The concept of Smart Cities, has become the focal point in the global development agenda. It is anchored through Smart Mobility, based on the premise of creating seamless movement within cities or towns using modal and multi-modal integration to achieve functional and operational integration. Notably, South Africa has also sought to bring about Smart Mobility through a government initiative of innovative urban public transport systems. This is anchored through the Integrated Urban Public Transport Network (IRPTN) program which prioritizes High-Speed Rail and Bus Rapid Transit (BRT) systems. Evidently, there has been a strong focus on the effective and efficient functionality of each mode, with minimal effort made for multi-modal functional integration both institutionally and operationally. Thus, this chapter evaluates functional integration (institutionally and operationally) across cities in South Africa and Zimbabwe. Findings suggest that at least the billing systems which form the BRT systems across the three metropolitan cities are integrated, while the Gautrain is completely disintegrated from the BRT system. Evidently, this has compromised attainment of Smart Mobility principles of efficiency, accessibility, convenience, reliability, and affordability through the implementation of innovative urban public transport systems. Subsequently, this chapter recommends institutional and operational integration, where route planning, scheduling, route management, and billing systems will be managed through a single-window coordination approach in order to enhance attainment of Smart Mobility Principles.

Keywords Multi-modal integration · Institutional cooperation and operational integration · Route planning · Billing system

6.1 Introduction

Smart public transportation is a global solution to reduce the high percentage of private vehicle users on the roads. Implementation of integrated public transportation in the developed countries has shown a positive impact in attracting private vehicle users and increasing the percentage of public transportation commuting. Different systems of urban public transport service different routes which makes it difficult for some of the commuters to reach their destinations smoothly. Consequently, this leads

T. Gumbo et al., *Urban Public Transport Systems Innovation in the Fourth Industrial Revolution Era*, https://doi.org/10.1007/978-3-030-98717-6_6

to increased desire for the use of private vehicles by individuals, and the operations of the different public transport systems are not synchronized.

Functional integration refers to the combining of services provided by different transportation providers. The most common form of functional integration involves sharing a ticketing system to enable easier transfer from one mode to another. Such an arrangement is dependent on the political willingness of the transportation providers and the proximity of their infrastructure (bus/train stops) to one another. Scholars such as Wong et al. (2008) and Zhang et al. (2020) have also extended functional integration to the combination of timetables for different modes, as this would lead to easy transfer between these modes.

Over the years, functional integration has been amended to include elements of social integration as a tool to classify commuters and identify their commuting needs. This has resulted in the reduction of negative externalities such as noise and vibrations from public transportation (Hesselgren et al., 2020; Pisoni et al., 2019; Vascik et al., 2018). Social integration has also been promoted as a practical societal service designed to educate and inform both authorities and commuters on a variety of issues namely economic, social and educational. This has led to the systematic planning for functional integration for different commuter needs.

This chapter examines (i) institutional cooperation and operational integration of urban public transportation systems, (ii) the route planning and scheduling, and route management of urban public transportation, and (iii) the billing systems of innovative public transportation. Further, we show how the current route planning and scheduling, and multi-modal integration in innovative public transport in Gauteng province are implemented in individual, innovative public transport systems, specifically the Gautrain and the BRT. Additionally, we highlight how the institutional operations of integrated Gautrain and BRT systems could be established between the two Gauteng systems. This chapter highlights the feasibility of having a connected multi-modal network in the Gauteng province.

6.2 Multi-modal Integration

Multi-modal integration enables travelers to combine two or more forms of transport to complete their journey. In this factsheet, multi-modal integration refers to the incorporation of cycling with additional forms of transport. Nevertheless, identification of transport users who are 'vulnerable to exclusion' is not a simple prospect. Multiple factors impact upon how well different groups of travelers are served by transport networks. These include the person themselves, contextual factors such as the geographic and economic environment, and existing multi-modal transport systems (Mbatha & Gumbo, 2020). Compounding these influences are spatial and temporal dimensions of service, with considerations around how effectively the transport system can match users with their preferred destinations at times that are appropriate for their desired activities (Cottrill et al., 2020). The residential and employment populations and relative densities of geographic areas have been noted as some of the primary influencing characteristics on mobility needs, services, and challenges (Chen et al., 2008; Schwanen et al., 2004; Stead & Marshall, 2001).

According to the U.S. Department of Transport (2015), multi-modal access to public transportation accommodates the many ways public transportation users get to and from a public transportation center to access a service. Those methods include walking, bicycling, riding feeder public transportation systems (e.g., taking the bus to connect to commuter rail at a station), and driving. The idea is that provision of infrastructure and support services for multiple modes of public transportation will increase its use and result in health benefits. Specifically, when effectively integrated, bicycling and walking to public transportation helps advance various environmental, health, and congestion-mitigating benefits for communities. A successful integration between modes will likely increase the catchment area and subsequent use of public transportation, and the efficiency of public transportation by reducing the necessity of feeder bus services.

While in Harare and Bulawayo in Zimbabwe, the urban public transport market is currently dominated by kombis and a small proportion of pirate taxis. Mini-buses mainly operate from main bus termini within the CBD area, while pirate taxis operate along arterials throughout the city center within the vicinity of intersections. There is no direct competition between these two modes because pirate taxis offer intracity (shuttle) service to passengers, for example, connecting one terminal point to another within a 2–5 km radius, while the mini-bus taxi caters for suburban travel demand.

The feasibility of the current multi-modal integration is clearly demonstrated by the Gautrain system in the Gauteng province. The Gautrain has support from Gaubus, as it is a feeder mode collecting commuters from different locations, where the high-speed train cannot be accessed and transport these commuters to the Gautrain or vice versa. Further, the Mini-bus also services some of the routes that are not used by the Gaubus to transport commuters to the Gautrain or the Gaubus.

6.3 Institutional Cooperation and Operational Integration

Institutional integration in transport systems refers to the combined management, planning, and governance of mobility modes (Merkert & Beck, 2020). Institutional cooperation has been noted as important for total sector performance, for example for achieving sector-wide environmental targets (Hull, 2008). Consequently, the institutional environment in which cooperative agreements are reached across modes and beyond the urban and firm boundary context is important for successful public transport integration to compete more effectively with private vehicles on the road. Institutional integration in public transport removes the impacts of boundaries on the transport system, leading to its function as one system rather than being fragmented.

A significant integration measure at the institutional level in some jurisdictions has been the grouping of all transport agency functions and operations under one administrative roof. An example would be Transport for London where the agency directs operations to achieve outcomes as directed by government (Luk & Olszewski,

2003). Similarly (but slightly differently) German Ver-kehrsverbünde, or transport associations, operate with both operator and government agency input to plan and deliver better integrated public transport to users and effectively compete with private transport (Buehler et al., 2019).

According to Sørensen and Longva (2011), in the UK, a "Commission for Integrated Transport" was established as early as 1998, while government transport agencies in Sweden and Finland were merged to achieve coordination and coherence; these countries were therefore among the first to coordinate public transport modes. In recent transport plans, Swedish governments have argued for increased coordination within the transport sector horizontally as well as vertically so as to achieve well-integrated public transportation systems throughout different modes of public transport from different service providers. Coordination between different public transport operators should work effectively to form a well-integrated system and avoid direct conflicts. It can be noticed that the operations become more successful when there is fare integration, information dissemination, timetabling between different modes of public transport, agreement between the different organizations to avoid conflicts, and different routes serviced by different public transport systems.

In the cities of Johannesburg and Tshwane, the Gautrain system has good operational integration within its system, as railways and roads are well connected. Minibus services transport commuters from one station to the next to service the Gaubus and the Gautrain; the Gaubus penetrates inside the suburbs/urban areas to service commuters who rely on this system; while the Gautrain is used for longer trips. The operation is well-integrated, as some of the techniques such as timetabling, scheduling, and payment systems are integrated. Commuters switch smoothly from one mode to another and are aware of the public transport scheduling. On the other hand, the BRT system in each Metropolitan Municipality functions on its own, and a commuter using Rea Vaya in the City of Johannesburg cannot switch smoothly to A Re Yeng in the City of Tshwane. The reason for this is that there is no solid point of spatial integration between Rea Vaya in Johannesburg and A Re Yeng in Tshwane. Nevertheless, slight integration in this system exists in the fare collection operations, as one smartcard can be used for boarding in any BRT system in the country. The operations of these two innovative public transport systems are good; but as yet, there are no points of connections and the BRT system lacks smooth seamless traveling provided to its users. Therefore, it is essential to integrate operations between the BRT and Gautrain systems in the province, as the Gautrain system can support BRT with the points of connection for all public transport users. With this established, travel within the province has been very smooth. This could be the starting point for the integrated public transport network to be established in the other provinces of the country leading to an integrated public transport network in the entire Republic of South Africa.

6.4 Route Planning, Scheduling and Route Management

According to Lucas (2013), the planning phase is critical, as a type of a system can be referred to as part of planning and the design element is tied to a management plan as it will involve planning of routes and stops integrated with frequencies. Therefore, at the strategic level of planning of the public transport system, long-term stability, and high service quality is required so as to influence urban development and to create more sustainable transport patterns (Nielsen & Lange, 2008). Consequently, planning a public transportation system is a multi-objective problem that includes line planning, timetabling, and vehicle scheduling (Gattermann et al., 2016). For each of these planning stages, models are known, and advanced solution techniques exist. Some of the models focus on costs, others on passengers' convenience. Setting up a transportation system is usually done by optimizing each of these stages sequentially.

In some instances, the challenge is one of physical connectivity, as some transport modes do not enter into certain locations, and the problem in some cases is in the planning phases. Thus, for urban public transport to function well, there are certain elements that need to be considered. The current demand in the planning process must be identified, as well as the relationship between movements of different transport modes and environmental demands. The plan predicting the future travel demand must be formulated, and a recommendation made to fulfill challenges that might occur. Further, there should be an assessment as to whether the proposal provides maximum benefit to the community. Consequently, to address these needs, there should be monitoring of existing conditions such as forecasting future population and employment growth, and identifying the projected land uses in the region and major growth corridors. There should be identification of current and projected future transportation problems and analysis, through detailed planning studies, of the various transportation improvement strategies to address those needs.

Nevertheless, urban public transportation planning in South Africa and Zimbabwe has improved over the years as more strategies from around the world are adopted. Several transport modes have been implemented to improve the state of public transport. For example, route planning is significant when planning for introduction of a certain public transport mode or public transportation system. People travel to different geographical locations and require a system that can deliver them without complications. Hence, route planning is very critical in public transportations as commuters are attracted to the mode of public transport that is easily accessible to get them from their origin to destination. In reality, one mode of public transportation or public transport system cannot service all desired routes for all riders. This is one of the factors contributing to a high number of private vehicle owners, as it affords traveling convenience; commuters desire a public transportation system that is fast, safe, reliable with less delays. Notable examples include the Gautrain system and BRT system. There is a need for integration of their operations, as both of these innovative public transport systems do not serve the same routes which make it an ideal to have an integrated innovative public transport system. Gaubus services most suburban areas to the city, and the BRT system services most township locations.

In the central business districts, the Gaubus and the BRT system services different routes, and only have a few locations that are spatial connected.

The route planning of the BRT in the Gauteng province has reserved lanes to give commuters fast and direct access from townships/suburbs to the city and other locations which makes the system attractive. However, the BRT system does not service all routes, and some locations are not serviced by this system in the province. The Gautrain system was developed to service most of the suburban locations to the city centers but not all the routes in the province. Nevertheless, the system is very fast and reliable according to its schedule. Gaubus services are provided every 12 min during peak hours and every 20 min during off-peak hours subject to road and traffic conditions. The system is well-structured as it consists of different modes which include high-speed trains (Gautrain), buses (Gaubus), and midi-buses. A Gautrain connects all three Metropolitan Municipalities in the province, and the stations of the Gautrain are allocated in the economic opportunity areas. Gaubus services the suburban locations and transport commuters to Gautrain stations for switching between transport modes so as to reach any location or City within the province, with an additional advantage of access to the OR Tambo International Airport Gautrain station. Therefore, the integration of the two systems indicates that more routes can be serviced, and more commuters could be attracted as they can travel smoothly across the entire province. Therefore, as future development of both public transportation systems extensions is put in place, the province can have a good functional and well-structured integrated public transport network.

6.5 Billing and Ticketing System

Public transport interchange has assumed a significance in the new transport agenda and the movement toward seamless travel by public transport (Gleave, 1998). Central to the requirement of a seamless public transport journey is the need to reduce the costs associated with interchanges, both perceived and actual. These costs can influence the demand for public transport in terms of the effects that the interchange has on time spent waiting, time spent transferring between vehicles, and the attendant risks and inconveniences involved with this activity (Hine & Scott, 2000).

According to Olivkova (2017), the Tag-in method is the important element between the user and the interface. This term is associated with the tariff and conditions of transport used in the realization of fare collection; it points to the organization of getting in and out of the public transport vehicle, and the definition of paid transport space within the stations. It also defines a method of selling tickets and their validation, which means producing a valid travel document to authorized ticket inspection personnel. The check-in system must be primarily supported by appropriate distribution of tickets, by their suitably wide range, by a clearing system, by an information system, and by the benefits of fare collection system for individual transport operators, etc. The check-in system directly affects how the passengers perceive the transport system and its services and are reflected in the final quality of the service which can be detected in a survey of passenger satisfaction or other suitable methods. Optimal setting of these factors can bring improvement of urban public transport services and thereby can positively contribute to the increase of new passengers at the expense of private car traffic which is one of the fundamental goals of integration in public transport.

A common global approach in the development of integrated multi-modal transport systems has been fare and ticketing system integration (Garcia & Azan, 2005; Luk & Olszewski, 2003). Fare system integration has been shown to facilitate "seamless" transfers and thus encourage the use of transfer routes (Sharaby & Shiftan, 2012). Integrated ticketing systems must satisfy two main requirements to become integrated (a) no additional cost for transfers, and (b) all modes and services to use the same ticketing system (Sharaby & Shiftan, 2012). Smart cards have been widely used to enable fare and ticketing integration, as they allow users to efficiently board and egress any type of service (Grotenhuis et al., 2007; Luk & Olszewski, 2003). Numerous European countries use integrated payment systems. In Belgium, for example, Brussels's STIB offers integrated pricing and ticketing, while Bilbao's Creditrans is an integrated ticket; Helsinki public transport pricing is an integrated fare system. It is noticed that the BRT ticketing system in the Gauteng province is integrated, as the smart card that is used for payment on A Re Yeng (City of Tshwane), and Rea Vaya (City of Johannesburg), and can be used for boarding Harambee bus (Ekurhuleni) or the other way round. However, the challenge is the spatial connection, as these modes are very far from each other spatially, and the trips are charged accordingly with no extra charges if a commuter is using Rea Vaya smart card to board in A Re Yeng. The billing system is a central point in connecting both Gautrain and BRT systems in the province, as spatially the three cities are connected by the Gautrain and BRT. Both service providers need to identify a common ground to make the operations intertwine, as alternatives are already feasible. While in Zimbabwe, commuters have the option to pay using cash or through the mobile phone-based money transfer platform commonly known as Ecocash. Although both of these forms of payment are accepted in mini-bus taxis and other modes, the challenge is that public transportation systems are yet to be integrated, hence each service provider charges the commuter separately.

6.6 Implications, Reflections, and Generalizability of Findings

Transport systems are complex and multi-dimensional with many parts that comprise the whole. The objective of integrated transport planning is to find balance among these dimensions so that planning and investment decisions contribute optimally to the economic, social, cultural, and physical potential of the transport system and society in general. Integration is a concern with the whole, with common objectives and agreed desired outcomes. Thus, different options, goals, and points of view must be combined to identify realistic solutions to community problems; thus, integrated transport planning is more than coordinated transport planning. It brings together multiple and sometimes-conflicting objectives to reach more sustainable transport outcomes that contribute to community, industry, and government priorities. Further, it is significant that a transportation network is well established from the beginning of the development; designs should allow roads to link and allow enough space for vehicles movement and for potential development of potential smart transportation.

Integrated public transport is a structural practice by which fundamentals of the commuter public transport systems (network and infrastructure, fares and ticketing systems, information, and marketing components) and a variety of carriers who serve different transportation modes, interact more closely and efficiently. It helps to generate an overall improvement in service quality level and enhanced performance of the combined public and individual transportation (Solecka & Zak, 2014). In general, the implementation of different transport integration solutions may result in decreased time spent traveling, public transport prices/costs, traffic jams, and environmental pollution.

According to Chowdhury and Ceder (2016), transfers are a key component in the successful operation of an integrated system. The roles of transfers are twofold. Firstly, transfers increase the accessibility of various destinations for users. Secondly, interchanges that facilitate transfers need to be provided in the network at strategic locations to reduce duplication of the public transport routes, thus improving the reliability of the network. Urban public transport integration solutions may improve the urban public transport network accessibility and competitiveness as well as assuring better utilization of different transportation means and infrastructures. The important contribution of an integrated urban public transport system is its ability to deliver to commuters a system that has alternatives; which does not limit desired travel routes; and which is appropriate, accessible, efficient, safe, effective, and reliable (Chowdhury et al., 2017; Ibrahim, 2003; Luk & Olszewski, 2003; Ülengin et al., 2007).

An integrated urban public transport system can draw more users. For instance, in Singapore, where urban public transport use is considerably high at 60% of mode share, the government aim in 2003 was to increase the mode share to 75% through integration (Ibrahim, 2003). Furthermore, Matas (2004) examined the important rise of urban public transport use in Madrid, Spain from 1986 to 2004 to more than 40%, and attributed it to changes made to integration. Again, in the City of Cracow, Poland, integrated billing and ticketing systems, and integrated scheduling through timetables, transfer points, and riding times improved the smooth movement of public transport (Solecka & Zak, 2014). Therefore, the Gautrain province has the potential for integrated innovative public transportation from different service providers. Operations can be connected on one platform through billing systems and scheduling, as there is good spatial connectivity in all three Metropolitan Municipalities between Gautrain system and BRT system. For both South Africa and Zimbabwe, it is critical that the public transport policies and legislative frameworks emphasize the support of institutional cooperation and integrated operations with the form of incentive support from the government and departments such as roads and transport. This will create interest among different service providers for the formation of an integrated public transport network.

6.7 Conclusion

The Gautrain system structure is a practical multi-modal public transportation that works well in the province, which demonstrates that the enhancement and integration of provincial public transport are possible. However, having different service providers working together to produce an integrated public transport system is not a new development. The review indicates that the province has what it takes to have a world-class, integrated, and innovative public transport system. With the availability of spatial integration, the billing system and ticketing is the first critical aspect that could be useful in connecting the two or more systems. This is because individuals can use one ticket/smartcard to board in both systems to complete their journey within any city in the province as well as have seamless traveling in the province. Moreover, scheduling of public transportation operations indicates that multi-modal transport can work when structured well. Therefore, when the scheduling aspect of both systems can be planned to be compatible with one another, integration of these two systems can work well.

References

Buehler, R., Pucher, J., & Dümmler, O. (2019). Verkehrsverbund: The evolution and spread of fully integrated regional public transport in Germany, Austria, and Switzerland. *International Journal of Sustainable Transportation, 13*(1), 36–50.

Chen, C., Gong, H., & Paaswell, R. (2008). Role of the built environment on mode choice decisions: Additional evidence on the impact of density. *Transportation, 35*(3), 285–299.

Chowdhury, S., & Ceder, A. A. (2016). Users' willingness to ride an integrated public-transport service: A literature review. *Transport Policy, 48*, 183–195.

Chowdhury, S., Hadas, Y., Gonzalez, V. A., & Schot, B. (2017). Public transport users' and policymakers' perceptions of integrated public transport systems. *Transport Policy, 61*(2018), 75–83.

Cottrill, C. D., Brooke, S., Mulley, C., Nelson, J. D., & Wright, S. (2020). Can multi-modal integration provide enhanced public transport service provision to address the needs of vulnerable populations? *Research in Transportation Economics, 83*, 100954.

Garcia, M. M., & Azan, S. (2005). Integrated transport systems in Latin America. *Public Transport International, 54*(5).

Gattermann, P., Großmann, P., Nachtigall, K., & Schöbel, A. (2016). Integrating passengers' routes in periodic timetabling: A SAT approach. In *16th Workshop on Algorithmic Approaches for Transportation Modelling, Optimization, and Systems (ATMOS 2016)*. Schloss Dagstuhl-Leibniz-Zentrum fuer Informatik.

Gleave, S. D. (1998). *Leeds travel blending*. Final report, prepared for Leeds City Council and Metro. Steer Davies Gleave.

Grotenhuis, J. W., Wiegmans, B. W., & Rietveld, P. (2007). The desired quality of integrated multimodal travel information in public transport: Customer needs for time and effort savings. *Transport Policy, 14*(1), 27–38.
Hesselgren, M., Sjöman, M., & Pernestål, A. (2020). Understanding user practices in mobility service systems: Results from studying large scale corporate MaaS in practice. *Travel Behaviour and Society, 21*, 318–327.
Hine, J., & Scott, J. (2000). Seamless, accessible travel: Users' views of the public transport journey and interchange. *Transport Policy, 7*, 217–226.
Hull, A. (2008). Policy integration: What will it take to achieve more sustainable transport solutions in cities? *Transport Policy, 15*(2), 94–103.
Ibrahim, M. F. (2003). Improvements and integration of a public transport system: The case of Singapore. *Cities, 20*(3), 205–216.
Lucas, K. (2013). Transport and social exclusion. In *The SAGE handbook of transport studies*. SAGE Publications Ltd.
Luk, J., & Olszewski, P. (2003). Integrated public transport in Singapore and Hong Kong. *Road & Transport Research, 12*(4), 41.
Matas, A. (2004). Demand and revenue implications of an integrated public transport policy: The case of Madrid. *Transport Reviews, 24*(2), 195–217.
Mbatha, S., & Gumbo, T. (2020). Identifying policies and legislative frameworks to create integrated innovative public transport in Gauteng Province. In *Shaping urban change—Livable city regions for the 21st century. Proceedings of REAL CORP 2020, 25th International Conference on Urban Development, Regional Planning and Information Society* (pp. 431–436). CORP—Competence Center of Urban and Regional Planning.
Merkert, R., & Beck, M. J. (2020). Can a strategy of integrated air-bus services create a value proposition for regional aviation management? *Transportation Research Part A: Policy and Practice, 132*, 527–539.
Nielsen, G., & Lange, T. (2008). *Network design for public transport success—Theory and examples*. Norwegian Ministry of Transport and Communications.
Olivkova, I. (2017). Comparison and evaluation of fare collection technologies in the public transport. In *16th Conference on Reliability and Statistics in Transportation and Communication, RelStat'2016* (Vol. 178, pp. 515–525), 19–22 October 2016, Riga, Latvia.
Pisoni, E., Christidis, P., Thunis, P., & Trombetti, M. (2019). Evaluating the impact of "Sustainable urban mobility plans" on urban background air quality. *Journal of Environmental Management, 231*, 249–255.
Schwanen, T., Dieleman, F. M., & Dijst, M. (2004). The impact of metropolitan structure on commute behavior in The Netherlands: A multilevel approach. *Growth and Change, 35*(3), 304–333.
Sharaby, N., & Shiftan, Y. (2012). The impact of fare integration on travel behavior and transit ridership. *Transport Policy, 21*, 63–70.
Solecka, K., & Zak, J. (2014). Integration of the urban public transportation system with the application of traffic simulation. *Transportation Research Procedia, 3*(2014), 259–268.
Sørensen, C. H., & Longva, F. (2011). Increased coordination in public transport—Which mechanisms are available? *Transport Policy, 18*(1), 117–125.
Stead, D., & Marshall, S. (2001). The relationships between urban form and travel patterns. An international review and evaluation. *European Journal of Transport and Infrastructure Research, 1*(2), 113–141.
Ülengin, F., Önsel, Ş, Topçu, Y. I., Aktaş, E., & Kabak, Ö. (2007). An integrated transportation decision support system for transportation policy decisions: The case of Turkey. *Transportation Research Part A: Policy and Practice, 41*(1), 80–97.
U.S. Department of Transport. (2015). *Multimodal access to public transportation*. U.S. Department of Transport.
Vascik, P. D., Hansman, R. J., & Dunn, N. S. (2018). Analysis of urban air mobility operational constraints. *Journal of Air Transportation, 26*(4), 133–146.

Wong, R. C., Yuen, T. W., Fung, K. W., & Leung, J. M. (2008). Optimizing timetable synchronization for rail mass transit. *Transportation Science, 42*(1), 57–69.
Zhang, H., Hu, Z., & Song, Y. (2020). Power and transport nexus: Routing electric vehicles to promote renewable power integration. *IEEE Transactions on Smart Grid, 11*(4), 3291–3301.

Chapter 7
Emerging Developments in Information Communication Technology and Commuting Trends in Johannesburg, Tshwane, Ekurhuleni, and Cape Town Cities

Abstract The use of technology has advanced operations in urban public transportation, and the introduction of the Fourth Industrial Revolution (4IR) can provide solutions for efficiency within the diverse services offered. There is no denying the role of technological innovation in enhancing citizens' livelihoods, which advocates for infrastructure which will upskill and generate opportunities in the 4IR. However, central to such technology practice is the development of techno-savvy people, to transform the economy and improve service delivery. In the transport sector, there exists legacy technology such as EMV (Europay, Master, Visa), but these have been disrupted by smart technologies solutions such as smartcard systems, online payment apps, smart stations, proving these are on the right path. Although technology has become pivotal, it is however not yet clear how this 4IR vision has brought the level of integration of public transport infrastructure closer to the sustainable and smart city concepts. In South Africa, specifically in Gauteng, interesting applications in 4IR solutions are notable in urban public transport systems such as the Gautrain but are skewed in other modes such as BRTs (Rea Vaya, A Re Yeng, and Harambe). Findings reveal that payments for the BRT system and the speed train system operating in the Gauteng Metropolitan Cities payments are conducted electronically, and information dissemination for both systems is also conducted electronically. Moreover, in the speed train system, there are machines providing self-service for both information distribution and payment methods. However, it has been noticed that there is no form of integration between the two systems. Therefore, the chapter recommends that both systems be integrated electronically, as this will create seamless traveling and more efficiency in the Gauteng urban public transport network by using one e-smart card. It will enable information scheduling connected with different systems allowing smooth switch in-betweens. In conclusion, despite the limited technological innovations in the transport sector, it is thus hoped that smart technologies solutions such as smart stations, smart ticketing, and biometrics solutions are applied to enable real-time data for traveling information.

Keywords Payment system · Information dissemination · 4IR · Urban public transport · Integration

T. Gumbo et al., *Urban Public Transport Systems Innovation in the Fourth Industrial Revolution Era*, https://doi.org/10.1007/978-3-030-98717-6_7

7.1 Introduction

Urban public transport assists of a large number of persons commuting daily in and out of the cities give them access to different activities anywhere and ensure that commuters are able to reach their destinations. User-friendly public transportation is required to make commuting easy and attract more private vehicle users. This can be achieved through integrated public transportation systems and use of Fourth Industrial Revolution (4IR) technological innovations implemented through information dissemination and payment methods. These are key components that can assist integration of different public transportation operations to improve the functionality of public transport. This is noticeable in developed countries with integrated, innovative public transportation systems which have implemented 4IR technologies to integrate public transportation electronically and which makes commuting convenient for commuters.

In South Africa, Gauteng province, innovative public transportation systems have been developed, but they are not integrated. The two innovative formal urban public transportation systems that have been developed are the Gautrain and Bus Rapid Transit (BRT) systems. These operate in the three Metropolitan Municipalities and are the BRT in the City of Tshwane known as A Re Yeng, the Rea Vaya in the City of Johannesburg, and the Harambee in Ekurhuleni. In addition, the Gautrain has high-speed trains, shuttles, and buses. The high-speed train railway line connects the three cities which are significant for both innovative public transportation systems to create seamless traveling, as these systems do not serve the same routes. There is a stigma that the Gautrain caters only for high and medium-earning classes, while the BRT system caters for medium and low-earners.

This chapter discusses the use of technological innovations in formal urban public transportation systems, specifically the Gautrain and BRT systems. The chapter also identifies the possibilities of integrating the two public transportation systems through combined electronic payment methods and information dissemination. Electronic operations in urban public transportation are some of the advancements brought forward by the 4IR to simplify the usage of public transportation in the Gauteng province for both commuters and service providers.

7.2 Technological Advancement in Urban Public Transport

The increasing development of big data technology has brought great opportunities and challenges for the innovation of complex systems such as Intelligent Transportation Systems (ITS) (Zeyu et al., 2017). The emergence of ITS has also paved the way for improving the safety, operation, and environmental impact of transport networks. Connected vehicles, a ground-breaking initiative of "intelligent vehicles," is emerging as the next wave of technology to further empower travelers (Jadaan et al., 2017). According to Mitchell and Suen (1998: 386), "Intelligent Transportation Systems is the term used to describe the application to road transport of advanced technologies including computing, sensors, communications, and controls." In accordance with the above, Intelligent Transportation Systems are designed and developed as a response to existing challenges related to the management of road traffic, pedestrian and passenger flow in large cities (Agureev et al., 2017). Further, Mandžuka et al. (2013) support the above statement by defining ITS as the application of information and communication technologies in the field of road transport. The usages include infrastructure, vehicles and users, traffic management and mobility management as well as interfaces with other modes of transport; these are all elements of 4IR technological innovations.

Developed countries use these technologies, and the number of applications has increased dramatically in the past years. For example, real-time display of information in public transport systems is becoming common, and multi-modal information terminals are starting to appear (Liu & Ceder, 2017). ITS technologies are playing an important role, as governments around the world are employing them for improving the functioning of transport systems resulting from challenges of increasing population and congestion in cities. This tool makes a dramatic difference in public transportation, and it could be applied in many ways with different models. One of these is the Advanced Traveler Information System (ATIS) which monitors the movements and location of a certain source of transport (car, bus, etc.) in real-time and watches the condition of the transport network (Briem et al., 2017).

In addition, Mitchell (1997) states that "The areas of application of ITS are usually listed as Advanced Traffic Management Systems (ATMS), Advanced Traveler Information Systems (ATIS), Advanced Vehicle Control and Safety Systems (AVCSS), Advanced Public Transport Systems (APTS), Emergency Management (EM), Advanced Rural Transportation Systems (ARTS) and Commercial Vehicle Operations (CVO)". The functioning of Intelligent Transport Systems improves the safety, efficiency, and capacity of the roads, monitors urban public transport users, and provides better information before and during a journey. This includes the use of e-smart cards to pay public transport fares, all of which reduce stress, increase security for travelers, and directly improve safety (De Grange et al., 2013). These technologies are deployed in the innovative urban public transportation systems implemented in the Gauteng province. An example of the technology which functions well in both systems is the Advanced Traveler Information System. This provides Public Information Display at the stations where the real-time information of the mode of public transportation is provided as the time changes if there are delays and also can be noticed inside the bus or train.

7.3 Information Dissemination and Scheduling

Information Dissemination is defined by Gkania and Dimitriou (2019) as an active distribution and the spreading of information of all kinds to the users, or those audiences that deserve it. In developing effective dissemination strategies, plans, and policies for public transport companies there is a need to understand the scope and characteristics of their current and potential users. Appropriate dissemination is thus a significant aspect in attaining user satisfaction and increasing usage (Koutsopoulos et al., 2019).

According to Pojani and Stead (2015), in the dissemination and utilisation processes five fundamental processes need to be analysed namely:

User: the potential user of the product to be disseminated. The user is the receiver of information which can only be considered effective if the user is able to successfully utilise the information received to maximum satisfaction.

Source: the agency, organisation or individual responsible for creating new knowledge or a product for conducting dissemination activities.

Content: the knowledge of the product itself; an example in public transport content consists of travel times, routes, fares and timetables.

Medium: ways in which knowledge is shared or the product described and packaged or transmitted, for example, SMS's, e-mails, mobile apps, public information displays and social media.

Context: the way the product or knowledge is developed and disseminated, including contextual factors related to the source, the user, the content and dissemination medium.

There are typically great reasons why associations choose to disseminate information, they are normally related but can be categorised to underline the motivation and significance of effective information dissemination.

According to the NCDDR the following categories are reasons for information dissemination:

Judgement: Information is disseminated with the expectation that individuals within an organisation will improve their knowledge and subsequently improve their judgements in future situations.

Awareness: Information is disseminated with a specific end goal to teach, clarify and advance an idea, procedure or standard. For instance, technical stipulations explaining systems, capabilities, instruction about alternative transport to avoid congested routes, notification of train delays, are all ways in which information is disseminated.

Response: information is often disseminated with the sole expectation that it will bring about some feedback that may require additional data to be created. Examples include advertising, questionnaires, market survey, etc.

Collaboration: Information is regularly disseminated in order to share knowledge and ways of communication. Examples incorporate workflow systems to encourage the flow of information between systems in order to accomplish a common purpose, e.g., control systems where probes may identify and transmit notices about specific events.

The information distributed to commuters by the RRT system and BRT system is provided both electronically and physically in their stations. However, accessing information electronically is convenient, as individuals do not have to be physically in a certain place to access it. Moreover, smartphone applications can assist them to make informed decisions about the available public transport mode. The Gautrain has an application which provides information about the travel times of the train. If this application could be used to give schedules for both Gautrain system and BRT system in one platform, it could improve efficiency of the Gauteng formal urban public transport. Payment systems and information distribution in the province are operated and managed technologically with the use of both ITS and big data. Therefore, if the public transport operating in the province could be integrated through the use of these technological innovations, the speed of transport could be improved, and delays reduced.

7.4 Information Communication Technology

There is no single universal definition of Information and Communication Technology (ICT), but the term is generally accepted to mean all devices, networking components, applications, and systems combined which allow people and organizations (i.e., businesses, non-profit agencies, governments, and criminal enterprises) to interact in the digital world (Azolin et al., 2020; Litescu et al., 2015). ICT applications thus provide information through innovative services relating to different modes of transport and traffic management which enables various users to be better informed and make safer, more coordinated, and smarter use of transport networks. The use of ICT guides informs commuters how a certain public transport system works, and this information is provided in a portable manner. The use of ICT in urban public transportation also helps to compile data on the speed of vehicles in different parts of the city, or different sections of the road network, the time taken to move from place to place, the distance traveled, the points at which the vehicle stops and for how long, and other important information for commuters (Litescu et al., 2015). This helps in getting an idea of the level of congestion at various times in the day and the choke points in the city which need attention. This also helps in planning routes and schedules on a more scientific basis. Further, it also helps in monitoring driver and vehicle behavior for corrective action and provides information on the exact time at which a vehicle reached a certain place and the route it followed. This allows contracts to be monitored and a verification of whether contract terms have been met.

ICT has also made it possible to locate a vehicle and track its movements from a central control station. A global positioning system (GPS) device is fixed to a motor vehicle and programmed to continuously communicate its location to a central control center (Moyo & Musakwa, 2016). This information can simultaneously be communicated through specially developed apps to individual mobile phones. Similarly, a GPS embedded on most smartphones allows the location of a person using that phone to be communicated to a central control center, thereby allowing an assessment of the level of crowding in any area. The ability to accurately track a motor vehicle has led to many possibilities that were not otherwise feasible (Vuchic, 2017). In accordance with the above, the Gautrain system has an operational smartphone application for commuters to be able to track the real-time information of the movement of when the bus/train will leave or arrive at the stations. This makes the planning of a trip to a certain place of interest easier and avoids commuters having to wait for the train or bus at the station rather than arriving a few minutes before. Further, commuters who use both train and bus services of the Gautrain system can switch between modes smoothly, as they have access to information of compatible times of the two modes. ICT offers various platforms for information dissemination. Commuters that do not have access to smartphone applications also have the alternative of being informed with real-time information through Public Information Display systems.

Public Information Display systems (PID's) are defined by Vuchic (2017) as an automated system for supplying users of public transport with information about the state of a public transport service through visual, voice, or other media. Public Information Displays are becoming very common in modern public transportation, and real-time information displays are prominent in the transportation sector. These systems show real-time information by providing features such as next departure times of trains and buses at stations and stops. Research has shown that this kind of information is appreciated by commuters, and the installation of PIDs is set to significantly improve traveler information and the quality of service. In evaluating PIDs, commuters often question their reliability and comprehensiveness (Cats & Loutos, 2016).

Information sharing in urban public transport is therefore very crucial and needs to be accurate, as ridership will escalate. However, information distribution of public transport that services commuters in South African cities mostly have fixed timetables at dedicated stations. This can become a challenge on some occasions when a certain mode does not arrive. Furthermore, informal urban public transport does not have timetables at all, and the operating system is very different from formal public transport operations. (i) passengers need to wait for a taxi to be full to leave the taxi rank (taxi station); (ii) a taxi leaves the rank with a certain number of people and will fill the taxi up along the way so that the next taxi in line can be filled with the certain number. This could cause time delays, as the taxi will travel slowly while trying to get more passengers by stopping at regular spots which is according to the taxi association agreement; (iii) taxis drive on the main roads from origin (township) to destination (city center) and commuters can board anywhere along the main roads without knowing when the taxi will arrive or whether it will arrive. Therefore, innovative formal urban public transport presents options for better commuting; namely, a comprehensive, and easy-to-use passenger travel guide which is critical to successful multi-modal travel. The signage at rail and bus stations should be properly designed to convey effective information to travelers (Ferreira et al., 2017). Information Technologies (IT) and Intelligent Transport Systems (ITS) can therefore play important roles in integrated transport and information integration in general. For example, major railway stations in Japan have very clear signs differentiating directions to the high-speed rail network, the intercity train's network, and the suburban/local trains network (Wagner & Zündorf, 2017). In addition, websites provide public transport users with information on the multi-modal transport options available and the related details. These are some of the lessons which could help the Gauteng urban public transportation advance to an integrated innovative public transport Gautrain and BRT system.

7.5 Payment Systems and Ticketing

According to Evans et al. (2015), e-Smartcard technologies offer the potential for bringing transport and citizen services onto a single smartcard platform, referred to henceforth as an integrated 'Transport- Citizen' (TranCit) card. The wider integration of different services has been a key aspiration within the public transport sector, as it has the potential to deliver numerous benefits which will ultimately make public transport more attractive to users (Giuliano et al., 2000). Smartcards are not a new creation but have existed for more than 40 decades. For example, the first integrated circuit smartcard patent was successfully filed in Germany by Jürgen Dethloff and Helmut Göttrup in 1969. Throughout the 1970s, the technology gradually developed, and by 1977, Motorola and Bull had successfully developed a product that we would now consider to be a smartcard (Evans et al., 2015). E-Smartcards have become a regular feature of many peoples' daily lives, but they typically only provide access to a single service and are used in a restricted place, existing in isolation from other potential opportunities. The most outstanding examples of a single service smartcard are those deployed for public transport services (e.g., Oyster in London, Navigo in Paris, Octopus in Hong Kong). Bank cards are used for payments/cash withdrawal or in numerous organizations, where members use their smartcard as a proof of identification for building access control and as a bespoke key to other available services.

The idea of combining services onto a single card is not new, since major schemes have been experimenting with multi-application citizen cards since the late 1990s (Blythe, 1999; Blythe et al., 2000; Evans et al., 2015). Early integrations of the smartcard for public transport ticketing purposes were proven on a multi-operator or multi-modal basis (Giuliano et al., 2000). Other field operational trials of e-smartcard systems were favored by passengers and operators alike, saving time during boarding and being more convenient than cash-based transactions (Chira-Chavala & Coifman,

1996). However, in many instances, early attempts to develop integrated schemes failed due to the lack of a sustainable business case. These were a result of limitations of available technologies, no robust definition of the roles and responsibilities required for running a single system and a general lack of understanding surrounding the complexity of demands being made on the technology (Giuliano et al., 2000). Furthermore, e-smartcards are more convenient than paper tickets and paying hard cash to board a certain public transport mode. For example, public transport systems with electronic fare collection devices continuously store data related to trips taken by users. These contain valuable information for planning and policy analysis (Mbatha & Gumbo, 2019).

A single fare card for multiple transit services will facilitate the transfer between modes, and rebates can be implemented as an inducement for those who transfer from one mode to another, e.g., zonal rebates in Vancouver (Liu & Ceder, 2017). While electronic ticketing is not a prerequisite for integrated ticketing, it does provide a very powerful mechanism to operate an integrated fare structure efficiently and effectively. For example, Hong Kong, Singapore, and London all have smart card systems in place which has underpinned the increase of public transport usage (Ferreira et al., 2017). This accounts for approximately 85% of all main mode trips, respectively.

Some of the Gauteng province popular city buses such as the Metro bus have upgraded their fare collection mechanisms from paper tickets and coupons to e-cards (Makgoo, 2018). The e-card method in the province eliminated challenges leading to bus stoppages, which were a nuisance to commuters. Accordingly, the BRT system uses an e-smart card for fare collection, while the Gautrain system also uses an e-smart card known as a Gold card. The Gold card allows seamless transfers between Gautrain's train, bus (Gaubus), shuttles, and parking services. Once a commuter has purchased the gold card, one can use it to access and pay for either the train or the bus services. The MyCiti card also allows commuters to purchase goods at local shops as the card system is linked to a bank. This clearly indicates that MyCiti is moving toward electronic payments to resolve paper ticket, hard cash payment challenges.

7.6 Electronic Integration in Urban Public Transport

Urban public transport systems are complex, with many parts comprising the whole. The objective of integrated transport planning is to balance these dimensions so that planning and investment decisions contribute optimally to the economic, social, cultural, and physical potential of the transport system and society. Integration is concerned with the whole, with common objectives and agreed desired outcomes. The different options, goals, and points of view must be integrated to identify realistic solutions to community problems. Thus, integrated public transport planning is about more than coordinating transport. It integrates multiple and sometimes-conflicting objectives to reach more sustainable transport outcomes which contribute to community, industry, and government priorities (Mbatha & Gumbo, 2018). Further, it is significant that the public transport network is well established from the beginning of the development. Designs should allow roads to link, enough space for vehicle movement and for potential development of smart transport. In South Africa, the government implemented smart UPT such as the BRT system in the City of Tshwane (A Re Yeng bus), Ekurhuleni (Harambee), Cape Town (MyCiti), and City of Johannesburg (Rea Vaya bus). The development of this concept played a key role as it brought more urban public transport alternatives, however, it reduced the width of the roads for both public and private transport, leading to traffic congestion at certain peak hours. Hence, environmental challenges may arise and increase human health risks due to time spent on the road by cars releasing noxious gases. Further, the development of the Gautrain system which involves a train, bus, and shuttles has developed a railway line that is only used by high-speed trains, but the shuttles and buses use existing roads, and the system functions well.

Integrated public transport is a structural practice by which fundamentals of the commuter public transport systems (network and infrastructure, fares and ticketing systems, information and marketing components) and a variety of carriers who serve different transport modes, interact more closely and efficiently. This generates an overall improvement in service quality level and enhanced performance of the combined public and individual transport (Solecka & Żak, 2014). In general, the implementation of different transport integration solutions may result in decreased time spent traveling, reduced public transport prices/costs, traffic jams, and environmental pollution.

UPT should not be an independent phenomenon but should be viewed as dependent, as the prime objectives are efficiency, safety, reliability, and effectiveness. This means that UPT should connect modes of transport for necessary switching between them by commuters. However, transport-related innovations should not come only from the office of the transport department or experts in designing for current or future transport needs. The introduction of a unified system capable of connecting public transport would be difficult, however, it is possible. Additionally, there is an urgent need to reduce the number of transport challenges such as traffic congestion, air pollution, and car accidents, all caused by many cars on the roads. Moreover, the introduction of a good, integrated public transport system can attract many private vehicle users and increase the rate of public transportation commuting. Physical connectivity in South African UPT currently exists between the Gautrain system and BRT system, hence, the need for integrating the Gautrain and BRT systems electronically.

Further to the above, Chowdhury and Ceder (2016) state that transfers are a key component in the successful operation of an integrated system. The roles of transfers are twofold. Firstly, they increase the accessibility of various destinations for users. Secondly, interchanges that facilitate transfers need to be provided in the network at strategic locations in order to reduce duplication of the public transport routes and to improve the reliability of the network. Urban public transport integration solutions may also improve the UPT network accessibility and overall competitiveness as well as assure better utilization of different transport means and infrastructure. Integrated urban public transport also delivers a system with alternatives to commuters that do not limit desired travel routes and offer an appropriate, accessible, efficient, safe, effective and reliable system (Chowdhury et al., 2017; Ibrahim, 2003; Luk & Olszewski, 2003; Ülengin et al., 2007).

Some of the technologies enable 4IR technologies to function better such as big data and assist toward improving customer satisfaction. These include knowing transport times and delays that occur and assisting service providers to identify critical information such as knowing locations that need more capacity of services for commuters and identifying some of the challenges in the operations. Further, through the development of a single smart card for the Gautrain and the BRT systems, data collected can help to identify how the integration of the two systems works and what needs to be improved. Public transport must therefore be made more attractive and user-friendly in relation to improved service, travel information, reliability, safety, and the upgrade of infrastructure such as waiting stations (Adewumi & Allopi, 2013). Cost is another crucial factor influencing the demand for public transport in relation to the time spent waiting, boarding, and alighting from vehicles, coupled with the risks and inconveniences involved in those actions.

7.7 Integrated Public Transport Tickets

In April 2011, the Gauteng Department of Roads and Transport undertook the completion of the public transport systems planning and development of an Integrated and Interoperable Fare Management Framework. The objective of Integrated Fare Management (IFM) is to enable seamless travel and transfer across an entire journey using a single fare medium (and possibly a single fare) for different operators and modes of transport. Furthermore, it promotes an integrated fare collection system which will improve the transit experience and convenience for commuters. The IFM approach is intended to make public transport systems more efficient through better cash management and improving boarding times, which in turn reduces delays leading to better schedule adherence.

IFM relates to multiple operators (and multiple modes such as bus, rail, taxi) which deploy and accept the same fare collection mechanism for public transport services within a defined region. This type of system allows customers to travel throughout the region in a seamless manner. At a minimum, the use of a common fare medium permits commuters to load individual transit products onto a single card; (e-Tickets or passes) from multiple operators as well as e-Money (electronic cash for fare payment). The IFM Framework thus establishes a common basis and vision from which to promote and execute Integrated Fare Management and transport in Gauteng. The framework is developed on a provincial level (to ensure consistency for strategic components), although operational components of the framework will be carried out primarily at a municipal level.

7.8 Integrated Public Transport Information

The potential for collecting and integrating passenger travel information has always existed but has never been realized because complex surveys were necessary to record this data. However, this problem has been resolved with the advent of Integrated Fare Management and the requirement for fare collection systems to comply with National AFC Regulations through the use of bank-issued fare media. Many such electronic fare collection systems calculate and deduct a passenger's public transport fare through a "tap on tap-off" process. With the aid of vehicle satellite tracking systems, this data translates into information on passenger travel patterns and volumes. A central data warehouse collects and collates this passenger travel data from different operators. The data can then be analyzed to produce passenger travel information and real-time data for schedule information systems at stations and onboard buses. The advent of fast, reliable, and affordable wireless internet communication has made it possible for travel information to be disseminated timeously and reliably via mobile phone (SMS or social media), websites, and electronic signs. The use of integrated, bank-issued fare media must be established through compliance with National AFC Regulations, followed by an establishment of a Provincial Public Transport Data Warehouse. This will ensure the centralized collection and analysis of passenger data to be disseminated as required. A Passenger Information Call Center would be one means of coordinating the dissemination of such travel information.

The abovementioned highlights the province' desire to move toward integrated innovative public transportation through the use of technological innovations, as indicated in the Gauteng 25-Year Integrated Transport Master Plan. The technological innovations in the use of payment methods and information dissemination have the capabilities of establishing different public transport systems without any conflicts in the operations of each service provider. With proper measures considered by all involved stakeholders of the current implemented innovative public transportation system, it is possible to have an integrated system, as desired by the province.

7.9 Implications, Reflections, and Generalizability of Findings

Electronic integration is a key in developing functional operations and synchronized exchange of information in the innovative integrated urban public transportation services of the RRT and BRT. An integrated e-smartcard that can be used for payment in the two systems can make traveling easy for commuters using both systems to complete journeys or trips. This e-smartcard can make topping up or loading money easy, as any station can be used; for example, if the BRT station is not working or there are complications, a commuter can quickly go to the nearest Gautrain station to load money into the e-smartcard. An app that is used by the Gautrain system service provider can be advanced to include the BRT system services on one platform. This integration of information can assist commuters to plan trips and locate the real-time movement of the mode. It could further assist users who travel across the province from various locations where there are services of both systems. For example, a commuter from Soweto (City of Johannesburg) to Soshanguve (City of Tshwane) can take a BRT (*Rea Vaya*) to the city where there is a Gautrain station (Park station), take the Gautrain to the City of Tshwane and get off at the Pretoria station in the city, and take a BRT (*A Re Yeng*) to the final destination in Soshanguve (City of Tshwane). Another positive impact of the integrated mobile app information dissemination can be the reduction of time spent waiting for the mode of transport to arrive at the station. The times can be made compatible; for example, a bus from any location leaving at 6:00 am to any Gautrain station, and arriving at 6:40 am should have at least a ten-minute interval for a Gautrain/bus leaving at 6:50 am for commuters switching from the bus coming from these locations. This will create smooth traveling, can actually attract more users, and can make commuting faster for long or short traveling distances.

7.10 Conclusions

Innovative UPT, specifically the Gautrain and BRT systems in Gautrain have deployed the technological techniques used by some developed countries globally with world-class public transportation and have integrated public transport systems. The Gautrain system for information dissemination uses the Public Display Information system on platforms. Inside trains/buses, there are electronic machines that provide information regarding the functionality of the public transport system and a mobile app that commuters use at convenience spaces about the functionality of the public transport. Gautrain systems also use electronic smart cards for payment and there are self-service machines in the stations to load money onto the smart cards. Therefore, this system uses advanced technology for its operations and offers better and more convenient services for its commuters. It is an integrated public transport system in its own sense as it has shuttles, buses, and trains working together and servicing the three Metropolitan cities in the Gauteng province.

The BRT system has implemented the use of technological innovations in their operations as the information regarding the system such as the routes serviced, time of departure, and arrivals are distributed through BRT websites the Public Information Display system in the stations and inside the buses. Further, the payment system is managed through use of smart cards to tap in and out at the station terminals, and there are self-service machines installed for loading money into the smart cards, although these are not yet in operation. The BRT system is slowly improving its electronic services to improve the operations to make commuting easier. Further, some of the policies and legislative frameworks such as the Gauteng 25-Year Integrated Transport Master Plan in the province support the integration of urban public transport and the use of technology to make commuting convenient. Therefore, there are possibilities to integrate these systems electronically as they have both installed technological innovations and are supported by policies and legislative frameworks.

References

Adewumi, E, & Allopi, D. (2013). Rea Vaya: South Africa's first bus rapid transit system. *South African Journal of Science, 109*(7/8), Art. #a0029, 3 p. http://doi.org/10.1590/sajs.2013/a0029

Agureev, I., Elagin, M., Pyshnyi, V., & Khmelev, R. (2017). Methodology of substantiation of the city transport system structure and integration of intelligent elements into it. *Transportation Research Procedia, 20*, 8–13.

Azolin, L. G., da Silva, A. N. R., & Pinto, N. (2020). Incorporating public transport in a methodology for assessing resilience in urban mobility. *Transportation Research Part D: Transport and Environment, 85*, 102386.

Blythe, P. (1999). RFID for road tolling, road-use pricing and vehicle access control. In *IEE colloquium on RFID technology (Ref. No. 1999/123)* (pp. 8–1). IET.

Blythe, P. T., Lawther, S. J., & Shield, C. M. (2000). Using transport applications as the key to developing a business case for a citizen smart card. In *6th International Conference on Applications of Advanced Technologies in Transportation*. Newcastle University.

Briem, L., Buck, S., Ebhart, H., Mallig, N., Strasser, B., Vortisch, P., Wagner, D., & Zündorf, T. (2017). Efficient traffic assignment for public transit networks. In *16th International Symposium on Experimental Algorithms (SEA 2017)*. Schloss Dagstuhl-Leibniz-Zentrum fuer Informatik.

Cats, O., & Loutos, G. (2016). Real-time bus arrival information system: An empirical evaluation. *Journal of Intelligent Transportation Systems, 20*(2), 138–151.

Chira-Chavala, T., & Coifman, B. (1996). Effects of smart cards on transit operators. *Transportation Research Record, 1521*(1), 84–90.

Chowdhury, S., & Ceder, A. A. (2016). Users' willingness to ride an integrated public-transport service: A literature review. *Transport Policy, 48*, 183–195.

Chowdhury, S., Hadas, Y., Gonzalez, V. A., & Schot, B. (2017). Public transport users' and policymakers' perceptions of integrated public transport systems. *Transport Policy, 61*(2018), 75–83.

De Grange, L., González, F., Muñoz, J. C., & Troncoso, R. (2013). Aggregate estimation of the price elasticity of demand for public transport in integrated fare systems: The case of Transantiago. *Transport Policy, 29*, 178–185.

Evans, G., Guo, A. W., Blythe, P., & Burden, M. (2015). Integrated smartcard solutions: Do people want one card for all their services? *Transportation Planning and Technology, 38*(5), 534–551.

Ferreira, M. C., Fontesz, T., Costa, V., Dias, T. G., Borges, J. L., & e Cunha, J. F. (2017). Evaluation of an integrated mobile payment, route planner and social network solution for public transport. *Transportation Research Procedia, 24*, 189–196.

Giuliano, G., Moore, J. E., & Golob, J. (2000). Integrated smart-card fare system: Results from field operational test. *Transportation Research Record, 1735*(1), 138–146.

Gkania, V., & Dimitriou, L. (2019). A back-engineering approach to explore human mobility patterns across megacities using online traffic maps. In *Mobility patterns, big data and transport analytics* (pp. 345–363). Elsevier.

Ibrahim, M. F. (2003). Improvements and integration of a public transport system: The case of Singapore. *Cities, 20*(3), 205–216.

Jadaan, K., Seater, S., & Abukhalil, Y. (2017). Connected vehicles: An innovative transport technology. *Procedia Engineering, 187*, 641–648.

Koutsopoulos, H. N., Ma, Z., Noursalehi, P., & Zhu, Y. (2019). Transit data analytics for planning, monitoring, control, and information. In *Mobility patterns, big data and transport analytics* (pp. 229–261). Elsevier.

Litescu, S., Viswanathan, V., Lees, M., Knoll, A., & Aydt, H. (2015). Information impact on transportation systems. *Journal of Computational Science, 9*, 88–93.

Liu, T., & Ceder, A. A. (2017). Integrated public transport timetable synchronization and vehicle scheduling with demand assignment: A bi-objective bi-level model using deficit function approach. *Transportation Research Procedia, 23*, 341–361.

Luk, J., & Olszewski, P. (2003). Integrated public transport in Singapore and Hong Kong. *Road & Transport Research, 12*(4), 41.

Makgoo. A. (2018). *How to buy a bus ticket for the new cashless paying system.* Available from: https://rekordeast.co.za. Accessed January 09, 2021.

Mandžuka, S., Žura, M., Horvat, B., Bićanić, D., & Mitsakis, E. (2013). Directives of the European Union on intelligent transport systems and their impact on the Republic of Croatia. *Promet-Traffic & Transportation, 25*(3), 273–283.

Mbatha, S., & Gumbo, T. (2018). Electronic integration of urban public transport through automated payment systems. In *Readying Africa for sustainable development: A capacity building and strategy discourse. Proceeding of the 5th International Conference on Development and Investment Strategies for Africa* (Vol. 5, pp. 314–325).

Mbatha, S., & Gumbo, T. (2019). Identifying the possibilities of integrating speed train and the bus rapid transit system through mobile payment and information dissemination. In *Is this the real world? Perfect smart cities vs. real emotional cities. Proceedings of the 24th International conference on Urban Planning and Regional Development in the Information Society* (Vol. 24, pp. 385–390).

Mitchell, C. G. B. (1997). *Intelligent Transportation Systems (ITS) applications for improving transportation for elderly and disabled travelers.* Transportation Development Centre, Transport Canada.

Mitchell, C. G. B., & Suen, S. L. (1998). Urban travel, intelligent transportation systems, and the safety of elderly and disabled travelers. *Journal of Urban Technology, 5*(1), 17–43.

Moyo, T., & Musakwa, W. (2016). Using crowdsourced data (Twitter & Facebook) to delineate the origin and destination of commuters of the Gautrain public transit system in South Africa. *ISPRS Annals of the Photogrammetry, Remote Sensing and Spatial Information Sciences, 3*, 143.

Pojani, D., & Stead, D. (2015). Going Dutch? The export of sustainable land-use and transport planning concepts from the Netherlands. *Urban Studies, 52*(9), 1558–1576.

Solecka, K., & Żak, J. (2014). Integration of the urban public transportation system with the application of traffic simulation. *Transportation Research Procedia, 3*, 259–268.

Ülengin, F., Önsel, Ş, Topçu, Y. I., Aktaş, E., & Kabak, Ö. (2007). An integrated transport decision support system for transport policy decisions: The case of Turkey. *Transport Research Part A: Policy and Practice, 41*(1), 80–97.

Vuchic, V. R. (2017). *Urban transit: Operations, planning, and economics.* Wiley.

Wagner, D., & Zündorf, T. (2017). Public transit routing with unrestricted walking. In *17th Workshop on Algorithmic Approaches for Transportation Modelling, Optimization, and Systems (ATMOS 2017).* Schloss Dagstuhl-Leibniz-Zentrum fuer Informatik.

Zeyu, J., Shuiping, Y., Mingduan, Z., Yongqiang, C., & Yi, L. (2017). Model study for intelligent transportation system with big data. *Procedia Computer Science, 107*, 418–426.

Part V
Socio-economic Considerations in Urban Public Transport Systems

Chapter 8
Measuring Economic Benefits of Innovative Urban Public Transportation in the Era of 4IR

Abstract Emphasis on capturing 'value-add' in the form of fostering economic benefits as trickle-down effects emanating from Public Transport Infrastructure Investments is becoming a critical consideration worldwide. This is particularly true in the Fourth Industrial era, as Public Transport Infrastructure has been identified as one of the critical sectors which foster economic growth such as improvement of business operations, formation of Small, Medium, and Micro Enterprises (SMMEs), employment creation, and improved access to economic opportunities. In consequence, this chapter examines the ability of Metropolitan Cities Public Transport Infrastructure Investments to create direct and indirect economic benefits as part of investment returns (trickle-down effects). Findings reveal that there is minimal concentrated effort to foster economic benefits in the planning and implementation of public transport infrastructure initiatives. Such lack of collective strategy formulation reveals how much must be done by public institutions responsible for public transport infrastructure rollout and those responsible for economic growth and development. Evidently, fostering economic benefits as part of public transport infrastructure investments still needs to be incorporated as a crucial component in the planning and implementation of public transport infrastructure projects. A Transversal management approach is recommended, where interdependent institutions and stakeholders collaboratively working together to foster economic benefits. This forms the core of Public Transport Infrastructure Investment initiatives.

Keywords Infrastructure investments · Economic benefits · Public transport · Collective strategy formulation · Transversal management

8.1 Introduction

South African spatial and socio-economic disparities are complex and interconnected, with apartheid and colonialism as the main contributors to the status quo. This chapter reflects the extent to which the researcher was able to study the research phenomenon consistent with its aim and objectives. As already deliberated throughout the book, this study has investigated the impact of public transport infrastructure investments in redressing past spatial imbalances and socio-economic

T. Gumbo et al., *Urban Public Transport Systems Innovation in the Fourth Industrial Revolution Era*, https://doi.org/10.1007/978-3-030-98717-6_8

disparities. The chapter further synthesizes the findings and policy implications of the said research phenomenon. Cross-references are made to international experiences and the existing body of knowledge discussed in this book, and a fusion of the research outcomes is made, considering the three results chapters. This focuses on the effectiveness of policy and legislative frameworks, the spatial configuration of innovative urban public transport systems, and the impact of innovative urban public transport systems on business growth and employment creation. Policy impacts from each of the three mentioned subthemes of the research phenomenon are used to derive conclusions, for each of which specific recommendations are made. Based on these conclusions and recommendations, the chapter then discusses the research phenomenon with the goal of creating a framework for an integrated approach to spatial integration (urban mobility) through innovative urban public transport systems and socio-economic transformation.

The book has explored infrastructure projects, to look beyond mere provision of physical infrastructure, through the adoption of innovative approaches incorporating the creation of vibrant urban spaces characterized by business growth and employment creation. The Collaborative Planning Process is advocated as the backbone of the policy interventions to attain holistic, favorable development outcomes, where spatial and socio-economic disparities are simultaneously addressed through investments in public transport infrastructure. The chapter concludes by developing a framework for an integrated approach to innovative urban public transport systems and socio-economic transformation. The impact of innovative urban public transport systems on business growth and employment creation within the cities of Johannesburg, Tshwane, and Ekurhuleni forms the core of discussion and analysis. The study findings are represented, analyzed, and interpreted along with the socio-economic transformation indicators of business growth and employment creation, consistent with the research phenomenon being perused in this chapter.

Data collected through administering of questionnaires informs most of the discussions and analysis in this chapter. These questionnaires are carefully constructed after field observations to determine principal factors influencing business growth, consistent with public transport infrastructure investments and their future planning endeavors. Careful consideration is given to whether such innovations contributed to socio-economic transformation, with specific reference to prioritization of previously disadvantaged communities and the creation of an enabling environment in which small-scale entrepreneurs and informal traders could thrive.

The potential for socio-economic opportunities is assessed on the two main stations (connector points) in major intersections and public transport nodes within the three metropolitan cities. Key issues identified are cross-referenced, with lessons learned from international experience to find synergies and develop better strategies without neglecting the local context. The chapter then looks at the impact of the public transport infrastructure investments on employment creation. The chapter concludes by providing a synopsis of the findings and prevailing circumstances on business growth and employment creation through investments in public transport infrastructure.

8.2 Innovative Urban Public Transport Systems and Socio-economic Transformation

Transport planning has evolved from a traditional to a rational, comprehensive approach through an integration of all modes of transport while also focusing on investment returns and positive spinoffs in economic growth and development (Goosen, 2016; Litman, 2014). According to Liu (2005), the transport sector is one of the most important in the search for healthy economic growth, translating to socio-economic transformation and poverty alleviation. In this regard, it is important that an integrated approach to decision-making is adopted by all stakeholders to achieve economically sustainable human settlements (Pardo et al., 2010). Such innovation gears creation of sustainable, efficient, and cost-effective urban public transport systems (Hrelja et al., 2013; Transport Research and Innovation Portal, 2013). This is observed when there is a strong emphasis on the role of urban transport systems as a critical tool to integrate all sectors of the society while contributing to economic transformation (OECD, 2009). It can only be achieved through an integrated approach, where all relevant city stakeholders contribute positively toward economic growth, and where a functional, efficient public transport system is created to strengthen competitiveness and growth of small-scale businesses thereby boosting the local economy (Rosenberg & Weiste, 2007).

Moreover, Rode and Floater (2014) argue that through an efficient innovative urban public transport system, access to goods and services by people is well facilitated while fostering better economic benefits through economies of scale and integration effects. International experiences reveal that innovative urban transport systems like the BRT have led to improved economic and employment opportunities in cities like Bogotá (Colombia), Rio de Janeiro (Brazil) as well as the Chinese towns of Guangzhou and Lanzhou (Babinard, 2012; Bocarejo & Tafur, 2013; Jones et al., 2014). The common trend among these cities is integration of public transport systems into a multi-modal transport network, and creation of cost-effective urban public transport systems, given the local context, while focusing on addressing socio-economic challenges.

8.3 Transport Integration for Socio-economic Transformation

The genesis of integration in transport planning can be traced back to Europe during the 20th century, which the rest of the world followed (Walter, 2012). When defining transport integration, one cannot neglect social, economic, and environmental considerations (Potter & Skinner, 2000). In the same way, Louw (2003) argues that integrated planning looks beyond the dynamics of a well-coordinated transport system. Similarly, Preston (2012) defines transport integration as the process through which the planning and delivery of elements of the transport system are brought together across modes, sectors, operators, and institutions so as to increase net social benefits. Integration in transport planning is measured through best practices, where various analytical methods are applied to optimize an efficient transport network resulting in improved consumer experience and accelerated economic growth and development (Divall & Hine, 2017). Notably, cities with well-integrated transport systems are most likely to have an improved Gross Domestic Product (GDP) accompanied by higher levels of productivity (Graham & Niekerk, 2014). Land-use patterns of a city or town also influence the travel pattern transport network for socio-economic activities in a city or town. In this regard, the colonially inherited, fragmented spatial form in developing countries is such that coordination of land-use planning and transport planning is required to mitigate the special imbalances of the past through an integrated public transport network catering to the socio-economic needs of the local people.

Accordingly, integration in public transport systems is concerned with finding the right balances of the physical infrastructure, and socio-economic transformation elements through identification of desired outcomes and common goals. This implies that a well-coordinated and comprehensive transport system can still disintegrate if it neglects social, economic, and environmental considerations. Mechanisms must therefore be developed to ensure that transport planning investments and decisions do not compromise the value of social, economic, and environmental aspects of development, but focus on realistic solutions to community problems. Boschetti et al. (2014) also noted that innovative urban public transport systems with strong focus on integration will create cities that are innovative, and where trade, tourism, commerce, services, and education prosper.

8.4 Socio-economic Transformation Through Public Transport Infrastructure

It has become evident that transport planning is evolving from a traditional to a rational, comprehensive integrated approach that considers all modes of public transport while also fostering investments returns and positive spinoffs in relation to socio-economic transformation (Litman, 2013; Mackie et al., 2012).To achieve these, innovative urban public transport systems can be used as a catalyst for creating economically viable urban centers through an integrated approach. Fostering an enabling environment for the growth of small-scale entrepreneurship and which translates to employment opportunities and growth of the local economy should therefore be crucial to government infrastructure investment programs. Since the dawn of the democratic dispensation in 1994. South Africa has prioritized socio-economic transformation through numerous policy pronunciations. These efforts have been centered around reducing unemployment, poverty alleviation, and bridging the inequality gap (Gumede, 2013). Public transport infrastructure investment consistent with the IRPTN initiative, and driven by the national government reduce unemployment, poverty, and the inequality gap as part of its social contribution (City of Tshwane, 2016). However, there seems to be a lack of innovative approaches to contribute to socio-economic transformation as part of investments in the public transport infrastructure. Evidence on the ground suggests that there has been limited progress in bridging the inequality gap and alleviating poverty.

The complex South African socio-economic disparities resulting from the history of the labor market and distribution of economic opportunities is that of an unequal society, where access to economic opportunities assumes a disintegrated apartheid spatial form. Long-term unemployment is the most prolific, unprecedented negative consequence born of the country's history of skewed distribution of economic opportunities and spatial fragmentation (Van Der Westhuizen, 2012). Innovative urban public transport systems can however form the core of the IRPTN program as part of government infrastructure investments and foster socio-economic transformation through employment creation and upliftment of the previously disadvantaged.

8.5 The State of Business Operations Since Implementation of Innovative Urban Public Transport Systems

The state of business operations is essential to socio-economic transformation, as it provides a platform for small-scale entrepreneurs to thrive and for employment creation. This is also an often-ignored component of the informal economy in formal planning processes, whereas it has evidently supported most families of the urban poor or previously disadvantaged communities (Brown & McGranahan, 2016). Evidently, too little, if anything at all is done at the planning stage to ensure that the implementation of innovative urban public transport goes beyond improving urban mobility to create an enabling environment for small-scale businesses to grow, boost the local economy, and create jobs.

Even Municipal regional spatial development frameworks focus mainly on densification, with no emphasis on creation of activity nodes at major intersections along densification corridors which will enable small-scale entrepreneurship growth and employment creation. Where there is minimal improvement in business operations and lack of opportunities for upcoming entrepreneurs, poverty, unemployment, and inequality will continue to rise, with more people trapped in poor living conditions. The current results reveal limited improvements in business operations which can be attributed to public transport infrastructure investments. Moreover, the informal economy has hardly benefited from the implementation of innovative urban public transport systems. Figure 8.1 depicts the effects of innovative urban public transport systems on the state of business operations. The informal economy and the meter taxi industry are said to have benefited less from the implementation of innovative urban public transport systems. At least 60% of the people indicated that the implementation of innovations has not contributed to strengthening the informal economy, while the same margin (60%) also felt that meter taxis have also not benefited much. Cleaning

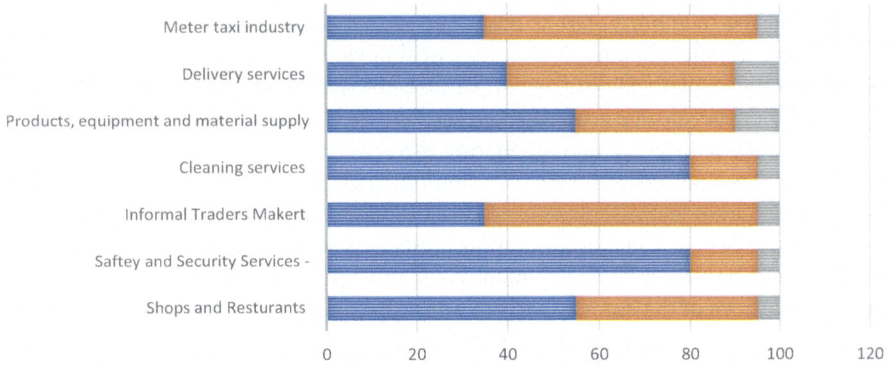

	Shops and Resturants	Saftey and Security Services -	Informal Traders Makert	Cleaning services	Products, equipment and material supply	Delivery services	Meter taxi industry
Yes	55	80	35	80	55	40	35
No	40	15	60	15	35	50	60
Not Sure	5	5	5	5	10	10	5

Fig. 8.1 Business operations improvement as a result of Gautrain and BRT bus services

services, safety, and security services, and products/material/equipment supplies were noted as the major business opportunities boosted and improved through the introduction of the innovative urban public transport system.

Notably, in relation to the sections on employment opportunities above, these types of jobs are dominated by the unskilled and semi-skilled labor market and low-income earners. The other types of business that are also said to have reasonably gained from the implementation of innovative urban public transport systems include shops and restaurants as well as products, equipment, and material supplies, with 55% of the people indicating these sectors as also having benefited. The material and equipment supply can be closely associated with the supply of cleaning material and related products. Therefore, it is not surprising that 50% of the people indicated that delivery service businesses have also gained from public transport innovations. It is evident that limited, though commendable improvements accrue with regard to business operations. Nevertheless, despite some of the businesses have benefited from the implementation of innovative urban public transport systems, it was also important for the researcher to determine whether these were upcoming entrepreneurs or already established businesses.

8.6 Components Promoting Business Growth

Integration of residential areas and economic nodes will ensure affordable public transport systems for the urban poor, thus enabling people to participate actively in the economy (Gahlot, 2017). Often, the urban poor spend most of their hard-earned money on transport, as they use more than one mode of transport for one trip and pay multiple fares, thus perpetuating poverty and inequality (Franklin, 2014). People's ability to obtain easy access to economic and other urban opportunities at an affordable transport fare is a main contributing factor to socio-economic transformation, especially given the socio-economic needs and challenges of urban poor. This is because it eliminates the possibility of low-income earners who reside in isolated townships from becoming discouraged and abandoning their jobs due to spending most of their income on transport before they can take care of other basic household needs. Arguably, integration through innovative urban public transport systems improves levels of access to economic and other opportunities. However, integration of various urban opportunities is to a large extent dependent on the state of the urban public transport infrastructure and its ability to give local people access to available opportunities through affordable public transport, especially the urban poor. The need for human rehabilitation is evident, to enable people to become small-scale entrepreneurs alongside the creation of opportunities as part of public transport infrastructure investments. Thus, investment in human capacity and skills development should form an integral part of socio-economic transformation policy interventions. In assessing business growth within and around public transport infrastructure precincts, it was important to identify principal factors influencing growth. Such factors were identified by way of examining similarities and patterns consistent with the Chi-Square test for relational analysis. Consistent with the Principal Components Analysis, the following Principal Factors were identified using SPSS statistical software: Proximity to Public Transport nodes or Intersections, Business Location, and Easy Access. To determine the significance of business proximity to public transport nodes or intersections, the *P*-value of 0.05 was used as the significance level. The null hypothesis was that there is no relationship between business growth and proximity to public transport nodes or corridors, the alternative hypothesis being that there is a relationship between proximity to public transport corridors and business growth. For a significance level less than or equal to 0.05, the null hypothesis was rejected in favor of the alternative hypothesis (Table 8.1).

As depicted in Table 8.1, the *P*-value highlighted is less than 0.05, which clearly depicts a strong relationship between business growth and proximity to public transport infrastructure nodes. Thus, designing public transport nodes as sources for socio-economic transformation becomes a significant factor. The need for an integrated approach to public transport infrastructure and local economic development through promotion of small-scale business is thus supported as a significant feature that can boost business growth. Such should form the core principles and approaches to station designs, where stations are used as socio-economic hubs to boost the local economy.

To determine the significance of business location in relation to a high concentration of business activities and clients, the *P*-value of 0.05 was also used as the significance level. The null hypothesis was that there is no relationship between preferred business location and proximity to the market or clients and the alternative hypothesis is that there is a relationship between preferred business location and

Table 8.1 Significance level of proximity to public transport nodes or intersection

Chi-square tests

	Value	df	Asymptotic significance (2-sided)	Exact sig. (2-sided)	Exact sig. (1-sided)
Pearson chi-square	12.403[a]	1	0.000		
Continuity correction	10.601	1	0.001		
Likelihood ratio	11.504	1	0.001		
Fisher's exact test				0.001	0.001
Linear-by-linear association	12.279	1	0.000		
N of valid cases	100				

[a] 0 cells (0%) have expected count less than 5. The minimum expected count is 6, 38
[b] Computed only for a 2 × 2 table
Source SPSS analysis using Hypothesis Testing (*P*-value approach)

proximity to clients or market. For the significance level less than or equal to 0.05, the null hypothesis was rejected in favor of the alternative hypothesis (see Table 8.2).

Table 8.2 depicts that the significance level is exactly 0.005, which affirms that the null hypothesis must be rejected in favor of the alternative hypothesis that proximity to clients or market is significant in identifying business location. Thus, all the businesses which have the potential to thrive within public transport infrastructure precincts should be accommodated through a public transport node or intersection

Table 8.2 Significance level of business location

Chi-square tests

	Value	df	Asymptotic significance (2-sided)	Exact sig. (2-sided)	Exact sig. (1-sided)
Pearson chi-square	8.101[a]	1	0.004		
Continuity correction	6.915	1	0.009		
Likelihood ratio	8.389	1	0.004		
Fisher's exact test				0.005	0.004
Linear-by-linear association	8.020	1	0.005		
N of valid cases	100				

[a] 0 cells (0%) have expected count less than 5. The minimum expected count is 14, 57
[b] Computed only for a 2 × 2 table
Source SPSS analysis using Hypothesis Testing (*P*-value approach)

Table 8.3 Significance level of easy access

Chi-square tests

	Value	df	Asymptotic significance (2-sided)	Exact sig. (2-sided)	Exact sig. (1-sided)
Pearson chi-square	7.700[a]	1	0.006		
Continuity correction	6.500	1	0.011		
Likelihood ratio	8.349	1	0.004		
Fisher's exact test				0.007	0.004
Linear-by-linear association	7.623	1	0.006		
N of valid cases	100				

[a] 0 cells (0%) have expected count less than 5. The minimum expected count is 11, 16
[b] Computed only for a 2 × 2 table
Source SPSS analysis using Hypothesis Testing (*P*-value approach)

precinct plan. The significance of easy access to a business establishment in relation to the concentration of businesses within public transport nodes was determined using the *P*-value of significance level of 0.05. The null hypothesis was that there is no relationship between business accessibility and public transport structure accessibility, while the alternative hypothesis was that there is a relationship between business accessibility and public transport infrastructure accessibility. For the significance level less than or equal to 0.05, the null hypothesis was rejected in favor of the alternative hypothesis.

Table 8.3 depicts that the significance level is exactly 0.007 which clearly depicts that small business especially will grow and have more clientele if they are easily accessible through public transport infrastructure. The overall indication is that there is a close relationship between public transport infrastructure investments and business growth to the benefit of the local community within a city or town. Thus, all business establishments should be easily accessible through public transport infrastructure.

8.7 Place-Making in and Around Main Public Transport Stations

Looking at international experiences such as the City of Linz in Austria, it is evident that stations for innovative urban public transport systems can be designed as activity nodes for economic and social activities (Vougioukas et al., 2008). This researcher assessed through field operations how the stations are designed in the innovative urban public transport systems of the three Gauteng Metropolitan Cities, and whether they encourage vibrant social and economic activities in and around stations. While conducting field observations, the researcher also held informal discussions with people who were involved in and around the activity nodes or main stations, so as to assess the state as well as the potential of business operations in the stations' precincts. From the key informant interviews, it became evident that there is a limited effort to design stations as activity nodes for the socio-economic benefit of the local community. Rather more focus is given to densifying along the Reya Vaya, A Re Yeng, and Harambe BRT Corridors. This can be considered as an area of improvement in the future, to ensure that implementation of innovative urban public transport systems moves beyond improving mobility and connectivity, and becomes a catalyst for socio-economic transformation. The following sub-sections will give an overview of the researcher's field observations on the socio-economic activities in and around the main stations and major economic nodes visited.

The observations made during primary data collection suggest that there is a potential for small-scale businesses around the main stations and main intersections both within and outside major economic nodes. The significant aspect in the Gautrain station is the presence of established businesses and small-scale entrepreneurs (though very limited). This can be linked to the mixed-use (activity) nodes similar to the ones created as part of station designs in the City of Linz in Austria, as discussed under international experiences. There, stations can become nuclei for innovation and spatial structuring elements, to allow for vibrant socio-economic spaces to be planned and developed. Opportunities could be created for small-scale entrepreneurs, kiosks, or mini-shops providing perishable goods typical of those provided by informal traders in business nodes. The structure and physical appearance of these containers or trading structures should be in keeping with the aesthetic value around the station, economic node, and main intersection. Thus, an opportunity is available for socio-economic diversification and transformation by accommodating upcoming small-scale entrepreneurs. The concern evident in most key informant interviews was the need for human rehabilitation, as the challenge of dependency on the government has increased, with people unable to take full advantage of opportunities created.

It is accepted that informal trading has become a viable source of income for the urban poor and previously disadvantaged who have no fixed source of income or formal business (Adelzadeh et al., 2001; Srinivas, 2016). Given the amount of land evidently available that is often left underutilized, there is a potential for more innovative measures to be introduced for the diversification of the socio-economic space within major intersections, connector points, and main stations. This can also be used to accommodate informal traders (small-scale entrepreneurs) around the stations within innovative urban public transport systems. Such containers will allow informal traders to trade legally within structures that are in keeping with the aesthetic and spatial forms of economic nodes/intersections or connector points and can be vibrant socio-economic spaces.

In contrast, this is not the case in most intersections, main stations, and connector points. Even where a few economic activities are evident, they are not well-planned and effectively coordinated to attain socio-economic vibrancy, thus leaving room for improvement. Some BRT Service Stations are in close proximity to Gautrain stations and in some instances also in close proximity to Metrorail mass rapid train stations, and services of long-distance and local bus services as well as big taxi ranks. Evidently, the Main Stations, where all modes and types of public transport converge, are the ones where some elements of economic activities happen, at times in a less effective and efficiently managed manner.

Nonetheless, an opportunity is evidently present, where local economic development potential is created by concentration of these public transport activities in one node. Such a variety of public transport systems and operations concentrated in one location could be used as a mixed-zone activity node that can accommodate small-scale entrepreneurs, established businesses, and informal traders, given the different services they provide to the commuters and people working and living in the area. The Gauteng Cities should consider such initiatives in the form of a precinct plan linking main stations with City Centers, and where there are main taxi ranks and bus stations within a reasonable distance. Activities of small-scale entrepreneurs are evident, but with proper planning, these areas can even be transformed into a vibrant activity node for small businesses and socio-economic transformation to the benefit of the local economy.

8.8 Missed Opportunities in the Hatfield Stations Nodes

Within the City of Tshwane, unlike the Bosman station in Pretoria Central, there is not even a single shop in the precinct of the A Re Yeng and Gautrain stations in Hatfield. With such space, available as is evident in Fig. 8.2, this presents an opportunity to create an activity node. This will enable a vibrant opportunity for socio-economic transformation, where informal traders and small-scale entrepreneurs can trade and provide services to commuters and people living and working in the area, as both the stations are located within the economic node of Hatfield.

The picture on the top-left corner depicts the A Re Yeng Station, with the Gautrain bus stations nearby. While taking the picture on the top-left corner above, the researcher was standing right next to the Gautrain station. The picture on the bottom right corner depicts the Gautrain station taken from the same location as the one on the top-left corner which clearly illustrates the proximity of the Gautrain and A Re Yeng Stations. Considering the prospects that were already observed in the Pretoria Central Stations and comparing this with the socio-economic hubs created in the City of Linz train stations, the ample space available around the Hatfield station represents a missed opportunity for a vibrant economic node. Structures of aesthetic value like the one in Bosman/Pretoria Station in Fig. 8.3 could ideally accommodate informal traders in the Hatfield node.

The A Re Yeng and Gautrain stations are closer to each other in Hatfield than in Bosman (Pretoria Central). Though aspects of informal trading are not fully evident in the Bosman Gautrain station, at least a glimpse of it is evident, and a precedent has been set for a potential future expansion. This is hardly the case at the Hatfield station, where there is ample space that could be used for creating a mixed-use zone

Fig. 8.2 Spatial form around Gautrain and A Re Yeng stations in Hatfield. *Source* Author's illustration based on the primary data collection through questionnaires

Fig. 8.3 Spatial form around Gautrain and A Re Yeng stations in Pretoria Central. *Source* Author's illustration based on the primary data collection through questionnaires

or platform for socio-economic transformation. The Hatfield station thus presents an even greater opportunity for an innovative hub for small-scale business opportunities in the form of an activity node, with the two stations so close to each other. Such considerations can be made for other stations with further implementation of innovative urban public transport systems. This will translate into integrated efforts for socio-economic transformation and innovative urban public transport planning systems implementation.

8.9 Employment Creation Through Implementation of Innovative Urban Public Transport Systems

Pre-1994, the South African economy was a polarized labor market, where benefits of most skilled jobs were mainly accessible to the minority or whites, and the African people occupied low-income jobs in the unskilled or semi-skilled labor market (Leibbrandt et al., 2010). Redressing such imbalances occurred post-1994 through socio-economic transformation policy interventions (Triegaardt, 2006). Such investments have been channeled to improve the lives of the urban poor and to prioritize the previously disadvantaged for employment opportunities created. The major investments associated with innovative urban public transport systems create employment opportunities at both the construction and operational phases. The extent to which the previously disadvantaged and the urban poor benefited from such investment was one of the principal areas of investigation in the study.

In light of the pre-1994 circumstances and the country's priorities post-1994 (democratic) dispensation, the researcher assessed the impact of innovative urban public transport systems on employment creation and whether or not they reflect a transformed or transforming labor market from the pre-1994 polarized one. The prioritization of the previously disadvantaged and urban poor for employment opportunities was thus investigated through use of key informant interviews and administering of questionnaires. During this time, it became evident that most of the previously disadvantaged people lack the necessary skills, training, and expertise to take full advantage of some of the opportunities being created. For some who come from poor families, their background limits them from dreaming beyond their confined environment, i.e., isolated townships through the apartheid spatial form for human settlement development.

The only employment opportunities for the poor are as cleaners, security guards, drivers, cashiers, and queue marshals. As such, this perpetuates the polarization of the already skewed labor market, hence unemployment and inequality is constantly worsening and affecting the very people it affected the most pre-1994. While there are investments for infrastructure development, there seem to be limited investments to afford training opportunities for them to take full advantage of the opportunities created in the formal economy or skilled labor market. Thus, the imbalances of the past are still visible and thus leave much to be done to improve the effectiveness of the country's transformation agenda.

8.10 Construction Industry Employment Opportunities

It became evident during the key informant interviews that there are more semi-skilled and unskilled employment opportunities created during the construction phase of the public transportation infrastructure. This tends to be the main segment of the labor market that absorbs those from the previously marginalized racial groups, due to their limited skills and practices on-the-job training. However, employment opportunities that benefit the previously disadvantaged during the construction phase still perpetuate labor market polarization, as those who used to have low-income jobs in the unskilled labor market segment are still stuck in the same segment in the post-apartheid era. Most families are still left devastated or with no source of income when the construction phase is concluded. Nevertheless, this can be attributed to a lack of skills development and expertise in the skilled labor market, as they cannot be absorbed into the skilled labor market without the necessary skills.

As such, it becomes necessary to ensure that the major companies which are usually awarded the tender jobs are subjected, through a tender service level agreement contract, to give back to the local community through skills transfer for the youth. This will ensure that the gap is bridged, with more young people from the previously disadvantaged communities getting essential knowledge and expertise in the construction industry. This may also enable some to become future entrepreneurs in the construction sector, thus creating job opportunities for others. The trickle-down effects in the post-construction phase should contribute to improved business operations and small-scale entrepreneurship which will result in employment opportunities being created.

It should be noted though, that the construction phase job opportunities do not fully contribute to long-term socio-economic transformation, but rather provide a temporary basis for already struggling families. Though these opportunities make a difference in meeting essential household needs, they do not contribute much to socio-economic transformation on a long-term basis. Most of the jobs being created during the construction phase eventually end with time. Therefore, the need exists for encouraging Small, Medium, and Micro-sized Enterprises (SMMEs) along measured corridors and main stations of the innovative urban public transport systems, where more families will benefit meaningfully on a long-term basis. This is dealt with in detail later in this chapter when discussing the effects of innovative urban public transport systems on small-scale entrepreneurship and business operations.

8.11 Operational Phase Employment Opportunities

Though the labor market remains largely polarized, especially when considering the skilled labor market against the unskilled or semi-skilled labor market, there are employment opportunities being created during implementation of innovative urban public transport systems (see Fig. 8.2).

As evident in Fig. 8.4, most opportunities that have been created and have a strong presence of previously disadvantaged people include mostly the semi-skilled market. For general works and maintenance, security services, and operational services like drivers and/or Gautrain pilots, the people expressed 100% confidence that the previously disadvantaged are prioritized. A marginal 5% was less convinced that cleaning services prioritize the previously disadvantaged, but the overwhelming majority (95%) still indicated that cleaning services were one section of the labor market that the previously disadvantaged had benefited from. Though most of the jobs that the people were reflecting on included semi-skilled and unskilled labor, there are opportunities like those of Gautrain pilots, where reasonable transformation has been achieved. As already highlighted above, some of the opportunities created could not benefit the previously disadvantaged, because they lack the necessary skills and training to perform such duties. It is important to note that, where there were previously disadvantaged persons with the necessary skills and training to perform the function, employment opportunities were given to them. This is evidently true, as 70% of the people reflected that financial services employment opportunities have benefited those from previously marginalized backgrounds. Though these include cashiers as well, observations by the researcher were that there are management and senior management skilled job opportunities that are occupied by women and people from previously marginalized races.

Almost all the people who benefited in the skilled labor market opportunities and are from the previously disadvantaged racial groups have the necessary skills, qualifications, knowledge, and expertise to occupy those positions. These stressed the point already argued earlier that sometimes it is not a matter of prioritization but whether or not those who are supposed to be prioritized have the necessary

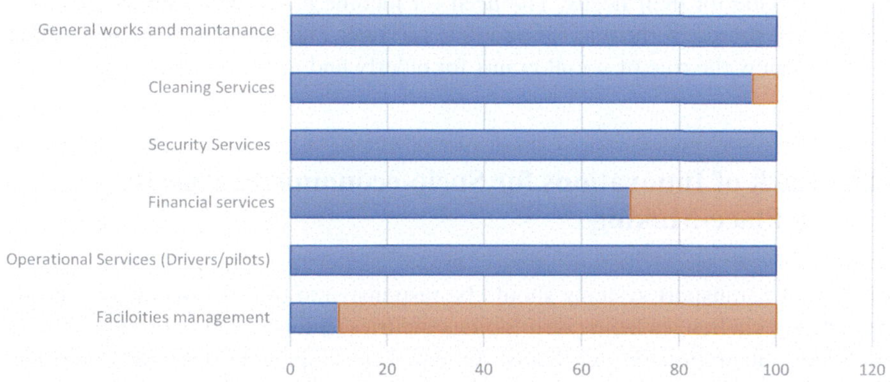

	Faciloities management	Operational Services (Drivers/pilots)	Financial services	Security Services	Cleaning Services	General works and maintanance
Yes	10	100	70	100	95	100
No	90	0	30	0	5	0

Fig. 8.4 Employment opportunities created

skills, training, and expertise to take full advantage of opportunities created. This still emphasizes the importance of investing in skills development and training as empowerment mechanisms for the previously disadvantaged to take full advantage of the opportunities created.

8.12 Prevailing Circumstances: Challenges and Opportunities

Social and economic transformation has undoubtedly been the core of South Africa's post-apartheid economic growth and development agenda (Leibbrandt et al., 2010). Nevertheless, spatial structures and forms inherited from the apartheid planning system have hindered social and economic transformation in most parts of the country (Turok, 2013). However, limited progress in the country's implementation of urban public transport planning and socio-economic transformation policies leaves much to be desired, despite the existence of good policy instruments and legislations. The defining feature has been the notion that issues of economic disparities and labor market polarization persist, with the previously disadvantaged being continuously marginalized. Ever-increasing levels of poverty and unemployment and the widening inequality gap continues, with more previously disadvantaged people trapped in poor living conditions due to a lack of economic means and systematic marginalization. To attain socio-economic transformation, an enabling environment for the growth of small-scale entrepreneurship and employment should be part of public transport infrastructure investments. However, questions have been raised as to whether the state has created a dependency culture, where people are solely dependent on government to provide for their needs. The need for income grants has been increasingly debated as a source of income for people's survival. This argument, though should not overshadow the role of social grants for elderly and poor communities.

8.13 Lack of Innovations for Socio-economic Prosperity in Place-Making

Urban public transport systems should be responsive to the socio-economic needs and complex challenges facing developing countries (Situma, 2002). In light of these difficulties, integration through Transit-Oriented Development (TOD) can contribute to better spatial forms and vibrant economic spaces. The lessons from the City of Linz in Austria discussed in Chap. 3 as part of international experiences show that an integrated approach to innovative urban public transport systems and socio-economic transformation creates sources for innovation, where stations or connector points can be designed as places of social and economic interaction. Thus, connector points or stations can be used as pivots for socio-economic innovation and prosperity by designing them as hubs. Here, small-scale entrepreneurs and informal traders can prosper and contribute to the local economy.

The study findings reveal that there is minimal opportunity for improvement in business operations and for up-and-coming entrepreneurs through the introduction of innovative urban public transport systems. Consequently, lessons must be drawn from the City of Linz to plan and design activity nodes where these are absent, in and around public transport precincts, so as to take advantage of missed opportunities to encourage small-scale entrepreneurs and the informal economy. Despite no signs of economic activities at the Hatfield economic node station, there were economic activities noted at the Pretoria Central economic node station, even though these are limited in quantity. This is an area that needs to be explored further and strengthened through precinct planning and design to create activity nodes to boost the local economy.

8.14 Unemployment and Poverty Trap

Through the implementation of innovative urban public transport systems, there are indications that employment opportunities are created. However, it was evident that the previously disadvantaged are still absorbed mostly into the low-income bracket occupied by the semi-skilled or unskilled. As such, income disparities in South Africa continue to reflect the historical imbalances manifested through a racial footprint and geographic location (van der Berg, 2010). Those who were previously marginalized continue to reside in isolated townships and have not escaped the low-income category. The state continues to produce policy instruments that are good on paper but achieve limited progress in terms of socio-economic transformation positively impacting people's lives. Thus, the ever-rising unemployment rate leaves much to be desired (Cilliers & Camp, 2013).

The reality is that those without formal education are subjected to low-income employment opportunities. These exist during start-up phase of construction projects, and can therefore not be considered as long-term sources of income. Though those with formal education benefited from the limited employment opportunities created, the majority were still systematically excluded because of their lack of the required experience. This stresses the need to invest in skills development for those without formal education. On-the-job training must be provided through government-subsidized internships and other interventions for those who hold essential qualifications in fields aligned to the country's investment priorities to enable them to take full advantage of opportunities created.

8.15 Toward a Framework for Socio-economic Transformation Through Innovative Public Transport Systems Innovations

Emanating from the three conclusions arrived at; it is evident that stakeholder's integration in policy intervention, cooperation by service providers for multi-modal networks, and creation of vibrant socio-economic spaces should always be the binding principles of public transport infrastructure investments. Notably, the said conclusions are interlinked, which necessitates an innovative approach to public transport infrastructure investments. Through the Collaborative Planning Process, issues associated with all three conclusions can be collectively addressed, as this allows for collective strategy formulation between interdependent stakeholders. Furthermore, implementation of innovative urban public transport systems with a strong focus on integration will enable cities to become nuclei for innovation, where trade, tourism, commerce, services, and education are improved. Of paramount importance is fostering investment returns and positive spinoffs through socio-economic benefits which lead to economic growth and development. This can be measured through easy access to economic and business opportunities, affordable transport options, and a fully functional, efficient, and integrated urban public transport network. Such a network should help reduce travel time and costs thereby improving productivity in workplaces. It will eliminate the element of unemployed residents' who are discouraged due to high travel costs for low-income earners (mostly from isolated townships), and who end up spending more than 50% of their wages on transport.

8.16 Principal Factors for an Integrated Approach to Innovative Urban Public Transport Systems and Socio-economic Transformation

The study advocates an integrated approach to innovative urban public transport systems and socio-economic transformation anchored through the Collaborative Planning Process. This emanates from the interdependent nature of the policy interventions, spatial connectivity, and socio-economic transformation. Nevertheless, as observed by Gil and Biezek (2016), the Collaborative Planning Process requires commitment and patient deliberation, as it brings together varied yet interdependent stakeholders through the common issues they deal with.

Prominent study findings about the effectiveness of policy and legislative frameworks, spatial configuration of innovative urban public transport systems, and business growth and employment creation can be summarised in the following principal factors:

- Status quo analysis
- Integration
- Urban growth management
- Inclusive growth
- Collective strategy formulation
- Decision making.

(a) **Status quo analysis**

The common issues reflected through the synthesis of the findings and conclusions affirm the need for understanding pre-existing conditions prior to development planning and implementation. A clear understanding of local knowledge will provide a solid base for stakeholders to identify issues and areas of prioritization.

(b) **Integration**

The need for integration cuts across the three complementary conclusions arrived at. For instance, Policy intervention relates to the need for stakeholders' integration or collaboration for innovative public transport systems. Spatial configuration relates to modal integration, while business growth and employment creation relate to integrating public transport infrastructure and local economic development investments.

(c) **Urban growth management**

For urban growth management, densification and a mixed-use approach to create self-sustainable communities were identified as essential. Establishment of activity nodes on major intersections along densification corridors was identified as essential for centralization of economic opportunities and promotion of small-scale entrepreneurship. Thus, vibrant urban nodes for social and economic interaction would enable people to live closer to economic opportunities or have easy access to opportunities.

(d) **Inclusive growth**

One of the common issues identified, given prevailing circumstances, is the need for strategies to ensure that the urban poor and previously disadvantaged communities are not systematically excluded, thereby indirectly perpetuating apartheid spatial and racial divisions. Affordable public transport should be a priority through an integrated billing system, where one's travel fares are charged per trip, regardless of the number of interchanges, as most urban poor often use more than one mode of transport to commute to economic opportunities. Human capacity development should be prioritized through investment in education, skills development, and human rehabilitation for previously disadvantaged communities to take full advantage of opportunities created.

(e) **Collective strategy formulation**

For an integrated approach to innovative urban public transport systems and socio-economic transformation, policies and implementation plans must be formulated collectively and agreed upon by all interdependent stakeholders. Such policies or development programs must ensure that urban mobility principles of accessibility, affordability, efficiency, and reliability are attained through innovative urban public transport systems investments. An enabling environment for business and employment growth should be created so that policy interventions and investments are responsive to spatial connectivity and socio-economic transformation development needs and challenges.

(f) **Decision making**

Due to the complex and interdependent nature of decision-making areas for public transport infrastructure and socio-economic transformation administered by various stakeholders, it is important that an investment decision by one stakeholder complements a decision by another. Designing an investment framework aligned to a collective strategy will assist in determining who will be responsible for funding which aspect of the holistic development program. This will guide stakeholders toward a common goal while eliminating the creation of 'white elephant' developments, where public funds are spent on development programs that are not responsive to spatial and socio-economic disparities.

8.17 Recommended Framework an Integrated Approach to Innovative Urban Public Transport Systems and Socio-economic Transformation

The Collaborative Planning Process forms the core of multi-disciplinary strategy development by interdependent institutions in terms of their core mandates and functions. Shakeri (2011) notes that the communicative platforms created for deliberation and engagement among stakeholders ensure that all contribute to devising a collective strategy. Thus, the Collaborative Planning Process is such that it must be robust and have well facilitated and trained planners so that the final outcomes yield positive socio-economic benefits. Figure 8.5 represents the framework for the recommended integrated approach to innovative urban public transport systems and socio-economic transformation anchored through the collaborative planning process.

To elaborate on the recommended framework, Table 8.4 is provided and is consistent with each principal factor based on the synthesis of the study. Key Performance Indicators (KPIs) were incorporated to monitor every aspect of the recommended framework.

Operationally integrating all (innovative and other) modes of public transport is encouraged as the backbone of spatial integration for effective functioning of

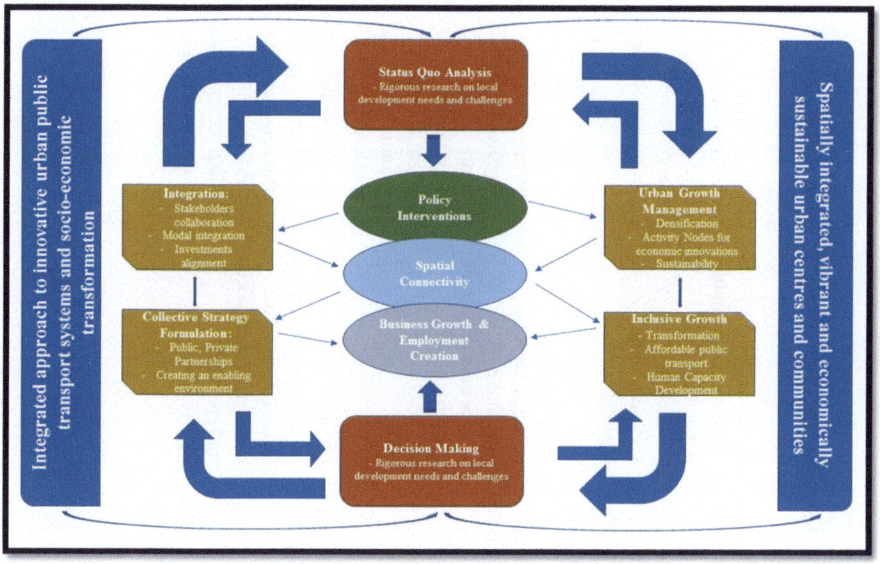

Fig. 8.5 Authors recommendation framework based on study findings

economic nodes and communities. When integrating varied modes of public transport, it is important to define the role and purpose of each mode or type of public transport in the overall public transport network to ensure improved urban mobility. For example, the Gautrain may be designated to connect major economic nodes, while other modes of public transport focus on commuting passengers between residential and economic or other urban opportunities. An integrated (multi-modal) public transport network will enhance spatial connectivity, densification, reduced travel times and costs, and efficiency and reliability of public transport systems to ensure synergies and a holistic approach to improved urban mobility.

Through modal integration, public transport infrastructure investments will be utilized more effectively to enable comprehensive public transport throughout municipal areas of jurisdiction instead of unnecessarily duplicating public transport services in one geographic area. A Public Transport Corporation should be created, where all public and private–public transport service providers collectively manage operations of an integrated (multi-modal) public transport network. This will enable centralization of operations and creation of integrated billing systems, with different stakeholders owning modes of public transport they originally operated prior to modal integration.

Government must create an enabling environment through legislative frameworks to allow for public–private partnerships. The policy formulation role should be separated from operational matters to eliminate state bureaucracy. Government should also not seek to replace the taxi industry or compete with it. Instead, it must be incorporated into the public transport corporation as an important stakeholder in a

Table 8.4 Recommended framework

Integrated approach to innovative urban public transport systems and socio-economic transformation					
Integrated approach to innovative urban public transport systems and socio-economic transformation	Principal factors	Policy and legislative frameworks	Improved urban mobility	Socio-economic transformation	Key Performance Indicators (KPIs)
	Status quo analysis	Rigorous research on pre-existing conditions on local development needs and challenges	Determining public transport infrastructure needs and challenges, types of operations, condition of different modes of public transport	Socio-economic profiling of the local community, public transport affordability, and socio-economic challenges	Identification of priority areas for policy intervention Identification of relevant stakeholders
	Integration	Integrating stakeholders with interdependent functional areas associated with identified development needs	Forging a working relationship for collaboration among public transport service providers Design for a single Integration public transport network	Integrated billing system charging fares per single trip regardless of interchanges Easy access to economic opportunities	Integrated public transport network covering every area Integrated billing system Integration of residential and economic opportunities
	Urban growth management	Government responsive to development needs by creating an enabling environment for public–private partnership without state bureaucracy	Housing opportunities for all income along public transport corridors, i.e., densification Public transport to help reduce traffic congestion and travel times	Creation of activity nodes mixed-use development around stations and connector points social and economic opportunities	Housing opportunities for all income groups along densification corridors Skilled professionals and entrepreneurs

(continued)

Table 8.4 (continued)

Integrated approach to innovative urban public transport systems and socio-economic transformation					
	Inclusive growth	Community-based planning where development planning is done "with" the people and not "for" the people Actively involve all interested and affected parties	Improved access to economic and other opportunities through affordable public transport Inclusion of mini-bus taxi industry in public transport upgrade and innovations	Human rehabilitation, skills development, and training for the previously disadvantaged urban poor in line with opportunities created	Affordable public transport An enabling environment for informal traders and small-scale entrepreneurs
	Collective strategy formulation	Planners as facilitators of the process responsible for conflict management and ensuring synergies among stakeholders	Provision of affordable, efficient, reliable public transport that enhances access to opportunities by redressing spatial imbalances	Development that encourages business growth, informal economy, and employment creation	Efficient, reliable public transport systems Business growth Employment creation
	Decision making	Development proposals formulated into two or more policy options for final decision making	Types and modes of public transport suitable for each segment of the community given their socio-economic profile and needs	Investments in social and local economic development Availability of funding for upcoming entrepreneurs	Implementation plan Investment framework and Monitoring and evaluation framework

Source Author's elaboration of the recommendation framework multi-modal integration. The state should refrain from autonomously imposing itself on the taxi industry by dictating terms through introduction of other modes of public transport modes in areas where taxis operate. Unlike government, the taxi industry has many cultural and organizational complexities and thus, negotiation should take place with all taxi owners to get full buy-in and cooperation instead of limiting discussions to taxi associations' leadership.

8.18 Conclusion

It is evident that South African spheres of government have not mastered the art of developing a collective strategy that sees their investments in public transport infrastructure complementing each other. The following recommendations reflect critical components for the above framework for an integrated approach to innovative urban public transport systems and socio-economic transformation. Synthesis of the findings, research conclusions, and lessons from international experiences which are relevant to local context has informed recommendations for the research phenomenon that was investigated.

This chapter has discussed the extent to which implementation of innovative urban public transport systems helped improve business growth and employment creation. Findings suggest limited improvement in business growth, especially in the integration of previously disadvantaged communities and informal traders to the local economy. Cleaning services, safety, and security services, and products/material/equipment supplies were noted as the core business opportunities that were boosted and improved. Even for the few opportunities created, limited entrepreneurial skills, knowledge, and expertise systematically excluded those from previously disadvantaged backgrounds from using such opportunities. Thus, readily established businesses are the main beneficiaries despite criteria for prioritizing BEE companies when awarding tenders.

Skewed attention to the public transport infrastructure rollout coupled with limited practical focus on trickle-down effects as socio-economic benefits was observed as another hindrance to the creation of vibrant socio-economic spaces through public transport infrastructure investments. Missed opportunities were observed, especially when it comes to accommodating the informal economy and small-scale entrepreneurs. The need to design main stations as activity nodes that encourage a variety of economic and social activities was identified as an area of improvement. Though limited in quantity, economic activities around the Pretoria Central precinct confirmed that this is indeed an area that needs to be explored further, while missed opportunities were noted in the Hatfield economic node.

A closer look was given to unemployment as it affects previously disadvantaged communities. Though employment opportunities were created, individuals from previously disadvantaged communities are still absorbed mostly into the low-income bracket occupied by the semi-skilled or unskilled. Lack of necessary skills sets, training and experience are contributing factors to the inability of those from previously disadvantaged communities to be absorbed into the middle and high-income categories. Where there were people from previously marginalized races with the necessary skills, qualifications, and expertise, they were prioritized. This stressed the need to invest in human capacity development for those from previously disadvantage communities to prepare them to take advantage of the opportunities created.

Human capacity and entrepreneurial skills development among the previously disadvantaged group is encouraged in order to enable people to take full advantage of opportunities created. Other than government tender business opportunities, small-scale formal and informal businesses can be encouraged through the creation of activity nodes for economic and social activities consistent with public transport infrastructure investments. If stations do not incorporate small-scale business opportunities, they become dull places with missed opportunities that could strengthen the local economy.

Government, in collaboration with other stakeholders, including academic institutions, should invest in human capacity development to ensure that people take advantage of opportunities created. Furthermore, government must also see to it that already established businesses mentor small businesses through mandatory skills transfer programs on government tender projects. There can be a service level agreement that each beneficiary of the government tender incorporates a skills transfer component to mentor and actively involve one or two upcoming entrepreneurs from previously disadvantaged communities. This will ensure that upcoming entrepreneurs acquire credible experience which will make them eligible to properly tender for future projects.

References

Adelzadeh, A., Alvillar, C., & Mather, C. (2001). Poverty alleviation, employment creation and sustainable livelihoods in South Africa.

Babinard, J. (2012). *A transport fare card moves Rio closer to social inclusion and carbon emission reductions.* World Bank: Transport for Development, Blog. Available from: http://blogs.worldbank.org/transport/a-transport-fare-card-moves-rio-closer-to-social-inclusion-and-carbon-emission-reductions. Accessed May 24, 2015.

Bocarejo, J. P., & Tafur, L. E. (2013). *Urban land use transformation driven by an innovative transportation project, Bogotá, Colombia.* Available from: http://www.unhabitat.org/grhs/2013. Accessed September 08, 2015.

Boschetti, F., Maurizi, I., & Cré, I. (2014). Innovative urban transport solutions. Available from: http://civitas.eu/sites/default/files/civitas-plus-innovative-urban-transport-solutions-www-final.pdf. Accessed October 04, 2017.

Brown, D., & McGranahan, G. (2016). The urban informal economy, local inclusion and achieving a global green transformation. *Habitat International, 53*, 97–105.

Cilliers, J., & Camp, H. (2013). Highway or byway? The National Development Plan 2030. *African Futures Paper, 6*, 1–16.

City of Tshwane. (2016). *Tabling of the city of Tshwane Integrated Development Plan for 2016–2021.* Available from: http://www.tshwane.gov.za/Documents/Tabling%20of%20the%

20City%20of%20Tshwane%20Integrated%20Development%20Plan%20for%202016%202021.pdf. Accessed June 18, 2017.
Divall, C., & Hine, J. (2017). *Transport policy: Learning lessons from history*. E-books.
Franklin, S. (2014). Job search, transport costs and youth unemployment: evidence from urban Ethiopia.
Gahlot, V. (2017). Integrated mass transit system in the Pink City: Mobility for All. *AQTR*. Available from: https://aqtr.com/association/actualites/integrated-mass-transit-system-pink-city-mobility-all. Accessed September 17, 2017.
Gil, N., & Biesek, G. (2016). Collective development of flexible capital designs: The case of future-proofing infrastructure. *IEEE Transactions on Engineering Management*.
Goosen, E. S. (2016). Transport—A catalyst for socio-economic growth and development opportunities to improve quality of life. In *Proceedings of the 35th Southern African Transport Conference (SATC 2016)*.
Graham, N., & van Niekerk. (2014). *Developing a public transport investment assessment framework*. SA Cities Network. Available from: http://www.sacities.net/wp-content/uploads/2014/10/sacn_public_transport_investment_assessment_framework_-web-.pdf. Accessed October 04, 2017.
Gumede, V. (2013). African economic renaissance as a paradigm for Africa's socio-economic development. In Kondlo, K. (Ed.), *Perspectives in thought leadership for Africa's renewal*. AISA Press.
Hrelja, R., Hansson, L., Richardson, T., Svensson, T., Lissandrello, E., Næss P., Tennøy, A., & Longva, F. (2013). *Innovations for sustainable public transport experiences and challenges in the Scandinavian countries*. VTI rapport 799A. www.vti.se/publications
Jones, P., Turner, D., & Heydecker, B. (2014). Congestion management for China's transit metropolis cities supported by the UK Foreign and Commonwealth Office prosperity fund.
Leibbrandt, M., Woolard, I., McEwen, H., & Koep, C. (2010). *Better employment to reduce inequality further in South Africa. Tackling inequalities in Brazil, China, India and South Africa. The role of labour market and social policies*. OECD Library.
Litman, T. (2013). Towards more comprehensive multi-modal transport evaluation. Best practices: Journeys, September 2013.
Litman, T. (2014). *Introduction to multi-modal transportation planning: Principles and practices*. Victoria Transport Policy Institute. Available from: http://www.vtpi.org/multimodal_planning.pdf. Accessed March 15, 2015.
Liu, Z. (2005). Transport investment, economic growth and poverty reduction.
Louw, J. (2003). Integrated transport planning: A Queensland experience. In *Proceedings of the 22nd Southern African Transport Conference (SATC2003)*. ISBN Number: 0-958-46096-5.
Mackie, P., Laird, J., & Johnson, D. (2012). *Buses and economic growth: Main report*. University of Leeds.
Organization for Economic Cooperation and Development. (2009). *Transport for a global economy: Challenges & opportunities in the downturn*. International Transport Forum: Policy Panel 1, Session 1: Transport and the Downturn Economic Impacts.
Pardo, F., Jiemian, Y., Hongyuan, Y., & Mohanty, C. R. (2010). *Shanghai manual—A guide for sustainable urban development in the 21st century*. Available from: https://sustainabledevelopment.un.org/index.php?page=view&type=400&nr=633&menu=1515. Accessed September 18, 2015.
Potter, S., & Skinner, M. J. (2000). On transport integration: A contribution to better understanding. *Futures, 32*(3–4), 275–287.
Preston, J. (2012). *Integration for seamless transport: International transport forum*. University of Southampton.
Rode, P., & Floater, G. (2014). Accessibility in cities: Transport and urban form. *The New Climate Economy, 3*.
Rosenberg, M., & Weiste, H. (2007). Medium-sized public-transport-city in the future.
Shakeri, M. (2011). Democratic assessment of collaborative planning practices.

Situma, L. (2002). Challenges of transport planning and spatial development in the Eastern Cape Province.

Srinivas, H. (2016). 14 reasons why the informal credit market is used by the poor: Policy implications for microcredit programmes in developing countries. Case Study Series E-111, April 2016.

Transport Research and Innovation Portal. (2013). *Innovation in urban mobility: Policy making and planning*. European Union.

Triegaardt, J. D. (2006). Poverty and inequality in South Africa: Policy considerations in an emerging democracy.

Turok, I. (2013). The resilience of South African cities a decade after local democracy. *Sage Journals: Environment and Planning, 46*(4), 749–769.

Van Der Berg, S. (2010). Current poverty and income distribution in the context of South African history.

Van Der Westhuizen, C. (2012). South Africa: Economic growth, poverty and inequality. *Foresight Africa: Top Priorities for the Continent in 2012*, 33–34.

Vougioukas, M., Sammer, G., Monzon, A., Evans, E., & Ambrosino, G. (2008). Quality public transport systems in medium size cities: Socio-economic development and urban regeneration impacts, funding and implementation advances.

Walter, R. (2012). Economic relations between Europe and the world: Dependence and interdependence.

Chapter 9
Societal Impact of Innovative Urban Public Transport Systems in South African and Zimbabwean Cities

Abstract Economic opportunity is often prioritized more than the human experience. Over the years this has led to a growth in the widespread belief that Urban Public Transport does not offer societal benefits to commuters (such as comfort and convenience). Hence, the proliferation of private vehicles in urban centers. Traditionally, urban public transport developers and institutions often aim at mastering the infrastructure developments and implementation of urban public transport. This trend has led to a knowledge gap of how recent advancements in public transportation in the era of the Fourth Industrial Revolution can be evaluated to ensure societal gains. This chapter, therefore, builds on the previous one by exploring the social impact of innovative urban public transport. A mixed-method approach was used to unpack case studies in South African and Zimbabwean cities. In recent years, crowdsourced data has become an integral element in assessing feedback or opinions from citizens and has provided reliable big data for the purpose of analyzing innovative urban public transport systems. Hence, GeoWeb 2.0 crowdsourced data was used to analyze the societal impact of public transportation systems across the metropolitan cities. Findings reveal that innovative urban public transport institutions have existing policies and corporate social responsibility functions. Furthermore, the cognitive elements influencing commuter mobility trends reveal commuter satisfaction rates. Thus, this represents a new platform to appraise existing measures for ensuring that the role of citizen-based innovation in public transportation is sustainable and is captured to public transportation development.

Keywords Societal · GeoWeb 2.0 · Crowdsourced data · Public transport · Commuters

9.1 Introduction

The global shift toward smart mobility has challenged the basis of traditional urban public transportation operations prevailing in much of Africa. Contemporary innovative urban public transport has been used as a tool for upscaling urban mobility (Ceder, 2020; Di Pasquale et al., 2016; Porru et al., 2020). Previous studies on urban mobility in African countries have further shown that economic opportunity is

T. Gumbo et al., *Urban Public Transport Systems Innovation in the Fourth Industrial Revolution Era*, https://doi.org/10.1007/978-3-030-98717-6_9

often prioritized more than the human experience (McKay et al., 2017; Musakwa & Gumbo, 2017; Ryseck, 2019). This has led to development of a two-tier system that informs infrastructure and services provision for the rich and the poor. For cities in South Africa, it has led to spatial variations in the level of availability of urban public transport infrastructure and services. Some spaces remain within proximity to economic opportunities in the form of transit-oriented development (TOD) while others are deprived. This spatial trend has led to a knowledge gap of how recent advancements in public transportation in the era of the Fourth Industrial revolution can ensure the societal gains regardless of economic classes. Literature has revealed crowd sourcing as a viable data collection technique, particularly during the Covid-19 pandemic, where social distancing is promoted (Marzano et al., 2019; Vermicelli et al., 2021). The analysis of such crowdsourced data can best be expressed through the insight of data analytics which has evolved over the years (Moyo & Musakwa, 2016).

This shift has taken place from seeing data as an end in itself to an enabler for decision making and collecting feedback for development (Chai et al., 2019; Mazumdar & Thakker, 2020). Furthermore, these analytical tools are a viable resource that will improve operational efficiency while boosting the quality of urban planning. This manner of data collection of commuter experiences has introduced new prospects for public transportation development, as it has revolutionized the speed and frequency at which information is gathered, processed, and evaluated. The South African government has also joined this advocacy for utilization of modern technologies by making a call for an Intelligent Transportation System (ITS) which is able to sense the prevailing conditions and react to them in real-time (Vanderschuren, 2006). While this is commendable, the dynamics of public transportation are still multifaceted.

Artificial Neural Networks (ANN) is a supervised learning system built of many simple elements, called neurons or perceptrons. Each neuron can make simple decisions and feeds those decisions to other neurons, organized in interconnected layers. Together, the neural network can emulate almost any function, and answer practically any question, given enough training samples and computing power. A "shallow" neural network has only one hidden layer of neurons while a Deep Neural Network (DNN) has a similar structure, but it has two or more "hidden layers" of neurons that process inputs. Both networks are effective for understanding high dimensionality problems, but they are also theoretically complex. The chapter will report the use of learning platforms to unpack the quality of service in public transportation systems. This proposed analysis is in line with several government policies and programs such as geared transportation and smart mobility. South Africa has adopted several enabling policies and legislative instruments to promote innovative urban public transport systems since the realization of the democratic dispensation in 1994. These commenced with the National Constitution of 1996, and the Green Paper on National Transport Policy launched in early 1996 which culminated in the adoption of the National Transport Policy White Paper later in the same year. Another important policy guide in the country is the National Development Plan (NDP).

The National Rail Policy Green paper was launched recently in 2015 to solicit views and ways of facilitating the planning and development of improved railway transport within the country. For example, The Gauteng province has implemented policies such as the Gauteng 25-year integrated master plan (ITMP 25), which it is envisioned will better the lives of Gauteng residents through the establishment of smart cities. Accordingly, the chapter promotes smart mobility and is aligned with the ITMP 25. South Africa is one of the few African countries which has made concerted efforts to improve their public transport systems. The economies and populations in South African cities are ever-growing, and efficient and effective urban public transport systems have become integral for the daily lives of South Africans. Hence, the Bus Rapid Transit (BRT) system and the Rapid Railway Systems (Gautrain) have been implemented since the beginning of the new millennium. This chapter seeks to build from the previous chapter by exploring the societal impact of innovative urban public transport.

9.2 Visualization of Points of Interest

This section presents the step-by-step approach of how the technical architecture of collecting and analyzing Web 2.0 data was undertaken using case studies from four metropolitan cities (City of Cape Town, City of Johannesburg, City of Tshwane, and City of Ekurhuleni) between 2020 and 2021. Crowdsourced data was collected from Web 2.0 platforms namely Facebook and Twitter for commuters across the four metropolitan cities to unpack the societal impact of innovative public transportation systems in these cities. The granularity of Web 2.0 data offers many merits for unpacking mobility research. For example, the digital footprint left by users allows researchers to quantify their socio-economic groups, but common limitations of Web 2.0 include data internet connectivity issues and the missing geographic coordinates of data. Furthermore, in developing countries, such as South Africa, mobile data costs are high, limiting the frequency with which commuters utilize social media platforms. Nevertheless, Fig. 8.4 reveals the demographics of respondents for the four metropolitan cities in South Africa. Given the socio-economic dynamics of the commuters of these cities, to assess connectivity for a public transportation, we draw from previous studies on the statistics of how communities have embraced the use of web 2.0 in their daily lives cities (Arnold et al., 2017; Huang & Wong, 2016; Jiang et al., 2019; Levy Abitbol et al., 2019) (Fig. 9.1).

Using Web 2.0 data of commuter perceptions on the operation's innovative public transportation, a query analysis was applied to filter social media posts to retain only the posts with the following key words Gautrain; Gaubus; Rea Vaya; A Re Yeng; MyCiti; BRT in South Africa and Mini-bus taxis for Harare and Bulawayo in Zimbabwe. An analysis of the Web 2.0 data further helped unpack the spatial influence of innovative public transportation based on the location of crowdsourced data.

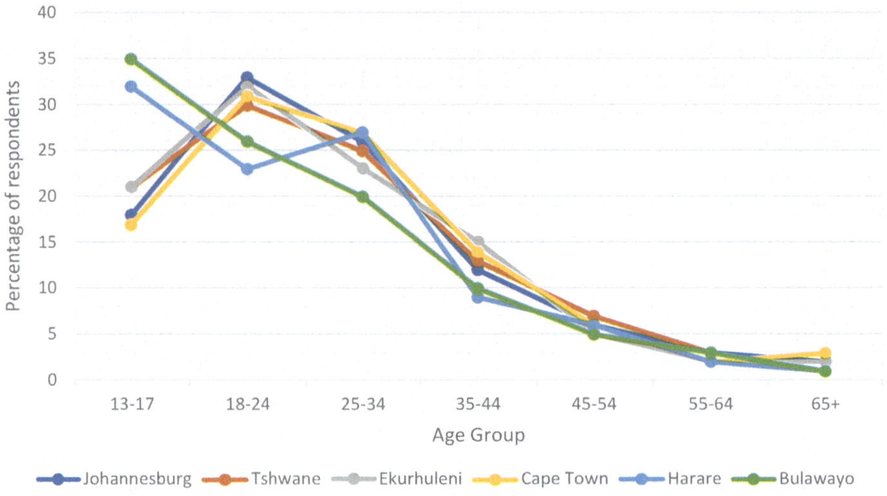

Fig. 9.1 Age distribution of Web 2.0 users

The population count of the Web 2.0 data after the query analysis was then subdivided as follows for visualization purposes:

- Blue representing 10 posts or less
- Yellow representing 100 posts or less
- Red representing 1000 posts or less
- Pink representing 10,000 posts or less
- Purple representing 100,000 posts or less.

Using four indicators the cognitive interpretation of commuters' experience when using UPT systems during various time of the day was assessed. The Web 2.0 data were then clustered using a neural network.

X is a cluster in which the posts are negative, $x(i)$ in relation to the respective indicator i, $x(t)$ is the number of elements of cluster X at time t.

Y is a cluster in which the posts are negative, $y(i)$ in relation to the respective indicator i, $y(t)$ is the number of elements of cluster Y at time t.

Z is a cluster in which the posts are neutral, $z(i)$ in relation to the respective indicator i, $z(t)$ is the number of elements of cluster Z at time t.

List of i indicators represented as:

a is an indicator for affordability relating to travel cost associated with UPT.

b is an indicator for travel time relating to average travel time associated with UPT.

c is an indicator for convenience and flexibility relating to soft clean seats, a pleasant temperature, preferably having air conditioning and not many people on the vehicle.

d is an indicator for Information relating to accessibility to travel information for trip planning.

The analysis then utilized Artificial Neural Networks to perform a linguistic measure analysis to understand the societal impact of innovative urban public transport (Gabdrakhmanova & Pilgun, 2021; Van Gerven & Bohte, 2017). Through the use of Web 2.0 data, these conceptions can reveal either positive or negative emotions, with the posts from 2020 to 2021 showing 5.6% and 3.9% emotions, respectively. These emotions can generally be defined as intense feelings or judgments made regarding a particular person or topic. The continued measurement of these negative and positive emotions is essential to develop innovative public transportation systems.

Consequently, the users' opinions about the innovative public transportation systems were grouped into the following linguistic measures;

Space—how much are people talking about the space around them when they refer to the innovative public transport systems.

Perception—how perceptive is the audience and are they open to greater interpretation through seeing and hearing or are they driven by a lower propensity of understanding.

Motion—words relating to movement, that is how long people are standing in queues. This also related to whether the queue moving or stationary.

Time—How much are factors of time being mentioned.

9.3 Discovering Functional Zones

At a regional level in South Africa (as shown in Fig. 9.2), the Gauteng province had the most users posting on social media about innovative public transportation. The Western Cape was the next province with the most users engaging on social media about operations of innovative public transportation. This could be due to the historical economic relationship between the two provinces, as they make the highest contribution to the country's gross domestic product. Also, these two provinces have had huge infrastructural developments to support the growth of innovative public

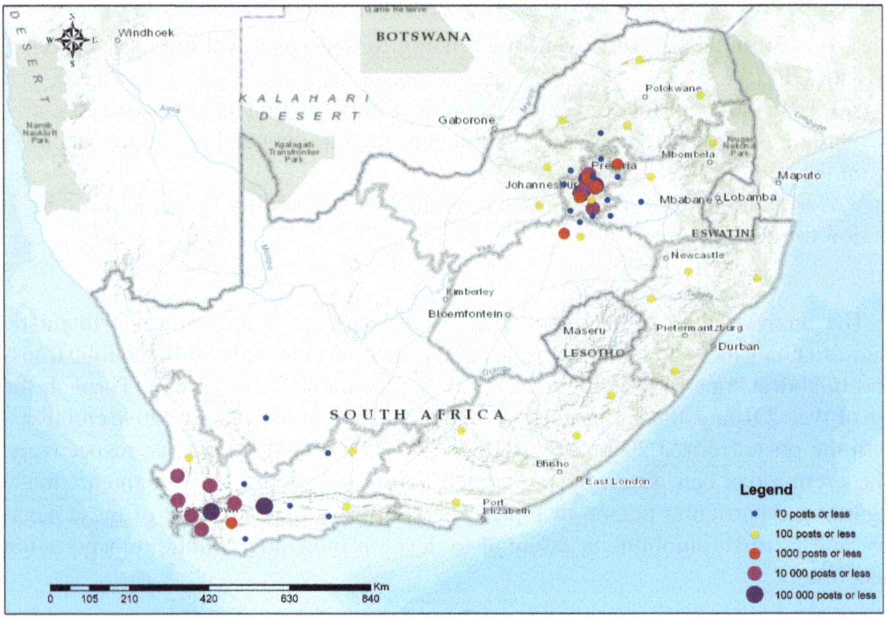

Fig. 9.2 Distribution of crowdsourced data in South Africa

transportation operations. Information from Zimbabwe reveals that the catchment area of the UPT is concentrated around the CBD of the two cities Bulawayo and Harare (see Fig. 9.3).

Figures 9.4, 9.5, 9.6, 9.7, 9.8 and 9.9 summarize results from the four indicators based on cognitive interpretation of commuters' experiences. Using indicators at city level the authors evaluate the cognitive interpretation of commuters' experiences as this provides a strong basis for defining the quality of service of public transportation that is easy to comprehend for the general public. Figures 9.4, 9.5, 9.6 and 9.7 reveal the level of affordability and commuting trip cost are essential for trips made during the morning and evening across all the cities. In South Africa approximately 62% and 52%. While in Zimbabwe for Harare 57% (see Fig. 9.8) and Bulawayo 48% (see Fig. 9.9). Most commuting trips during the peak hours are from home to work during the morning and work to home in the evening. Most commuters preferred routes with little traffic, hence these were marked as highly relevant in trip planning. In South Africa, 62% reported being anxious about traveling on roads with high speeding traffic, compared to 54% for cities in Zimbabwe. One of the factors considered to be of prime importance in the choice of routes is the existence of points of interest along the route. Generally, for both Zimbabwe and South Africa 50% of the participants prefer routes that pass along points of interest, while 16% do not choose routes along these and the remaining 34% are indifferent. These points of interest include parks, shopping areas, offices, and educational institutions.

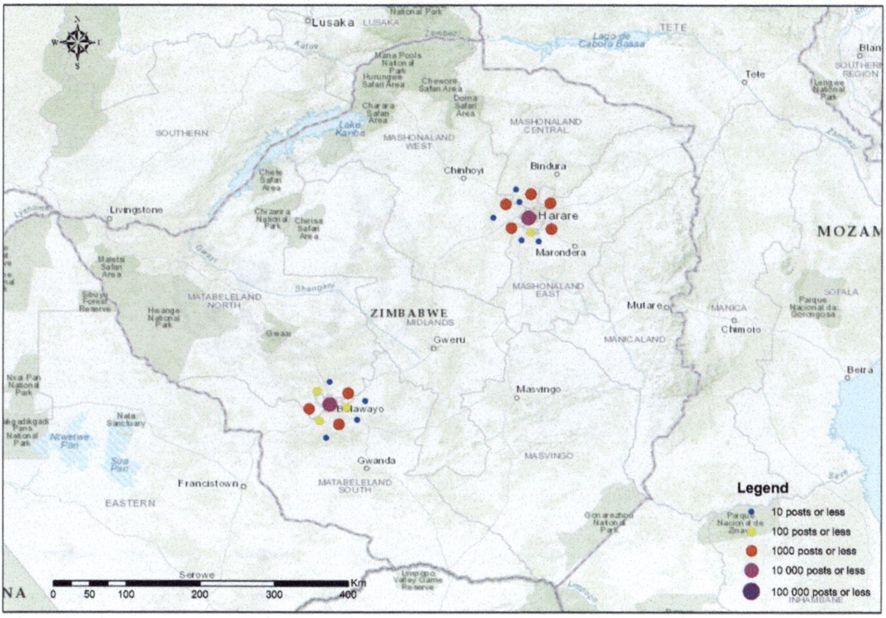

Fig. 9.3 Distribution of crowdsourced data in Zimbabwe

Fig. 9.4 City of Cape Town

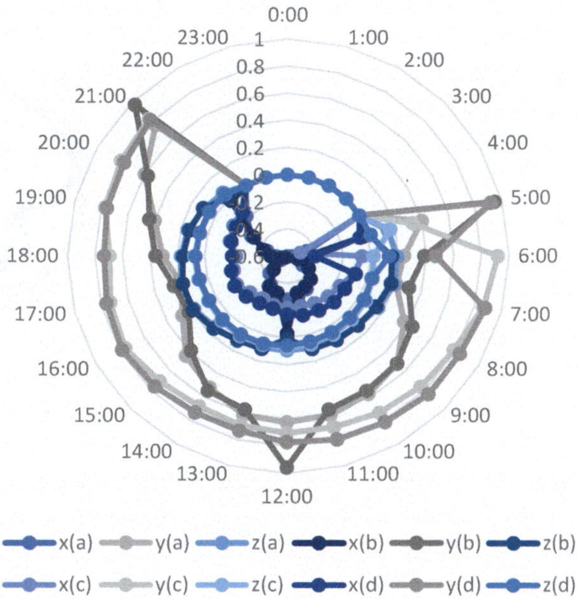

Fig. 9.5 City of Johannesburg

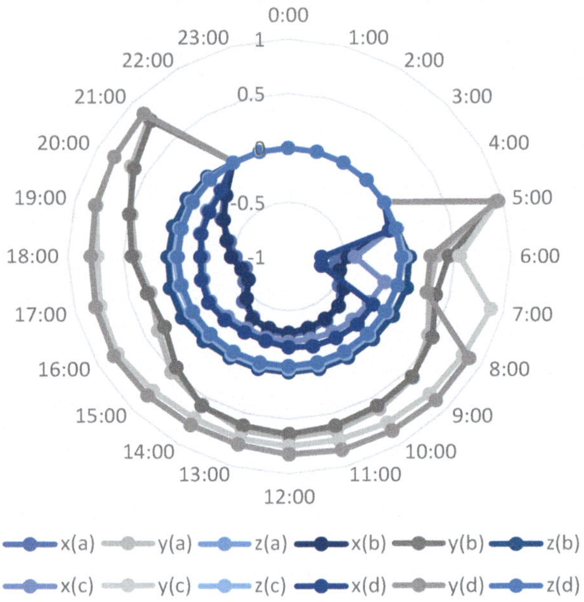

Fig. 9.6 City of Tshwane

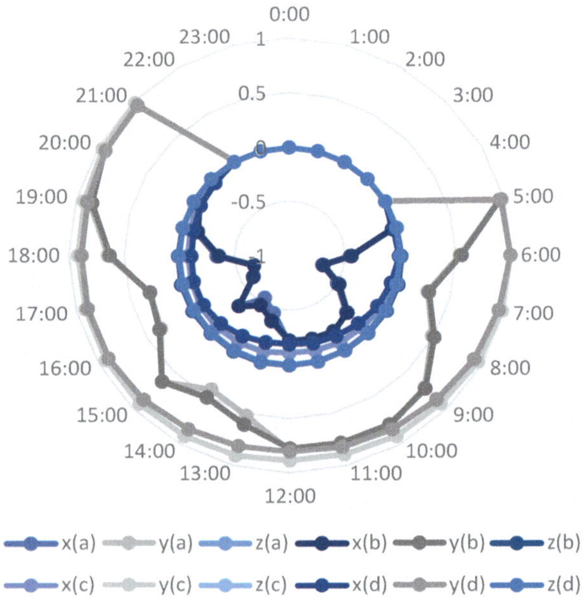

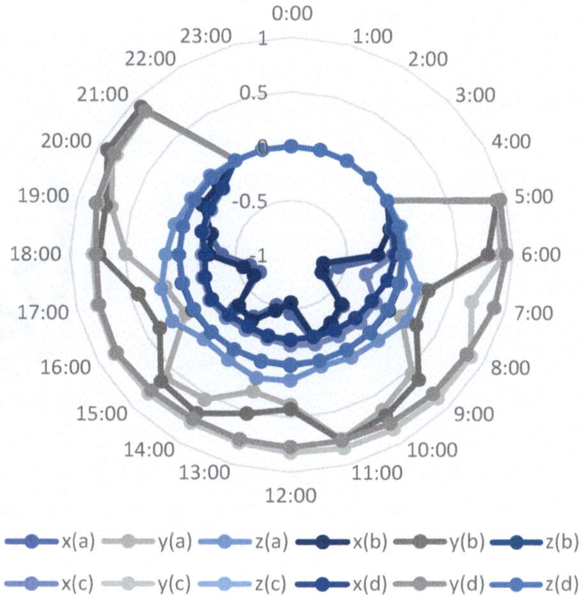

Fig. 9.7 City of Ekurhuleni

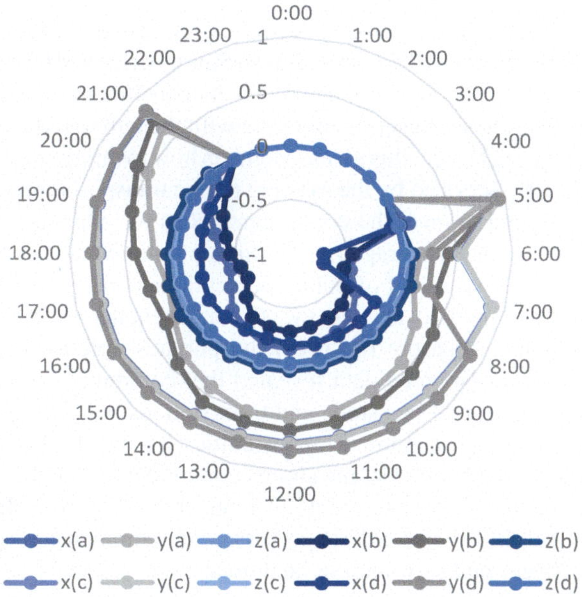

Fig. 9.8 City of Harare

Fig. 9.9 City of Bulawayo

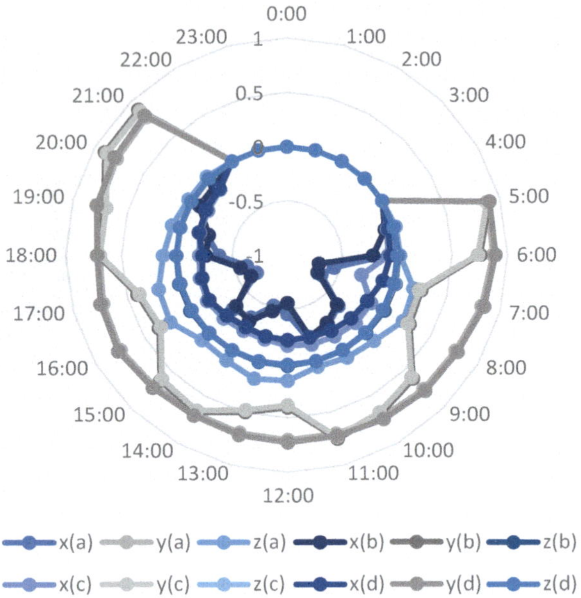

To a great extent, there seems to be a relationship with regard to trip length and trip duration, with 72% commuters in South Africa and 63% in Zimbabwe recording these as key factors in trip planning. It appears that most commuting trips are undertaken within the shortest possible time with the aim to reduce total distance traveled as the end goal. Hence, this gives insight on how to improve areas that are currently not being directly serviced by the existing public transportation infrastructure. They should be upgraded to meet the commuters needs for efficiency with regard to traveling time and distance coverage (i.e., promotion interchange stations).

The assessment of public perceptions about innovative public transportation has the potential to unpack the societal benefits of these modes. As pertains to the linguistic measure of space, the public's options were positive for all the cities. The crowdsourced data revealed that commuting trends were affected not only by the location of bus stops and train stations but also the routes of these modes, with routes along economic opportunities having the higher positive posts. Innovative public transport systems along economic nodes in the city would be able to connect commuters to these points of interest. This is also aligned with previous research which has shown a linkage between commuter perceptions with regard to service delivery and surrounding land uses.

The linguistic measure (see Fig. 9.10) also reveals another commuter priority for innovative reliable, fast, and efficient public transportation. The occurrence of delays within a system will lead to less favorable impressions being made. There seems to be a need for improvement with regard to time and the respective innovative public transportation systems, as the posts have generally been negative (low). Improving

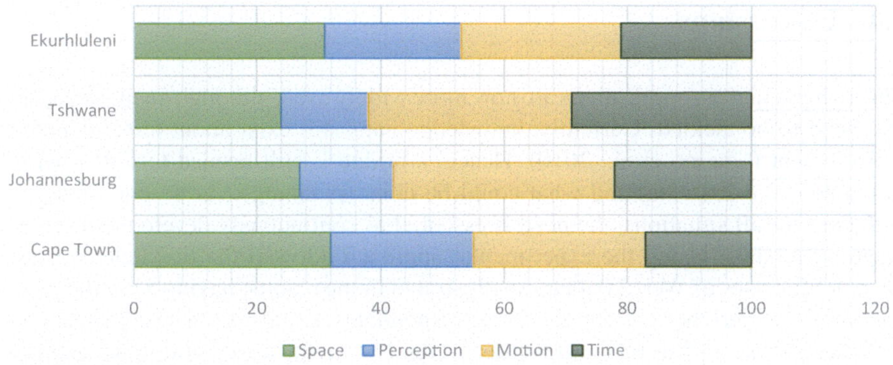

Fig. 9.10 Linguistic measure

the positive magnetism of innovative public transportation systems is essential for sustainable development.

Also, the manner in which commuters' relate to the linguistic measure of motion can either build or hinder the growth of public transportation. Studies have shown that excessive waiting seems to significantly hamper commuters perceived satisfaction. Lagune-Reutler et al. (2016) highlight how improving wait times perceptions is a principal issue with potentially lower costs than actions focusing on reducing actual wait times. With regard to the four cities' commuters, there are variations in perceptions. Generally, it seems that commuters' linguistic measure for motion is not constant in the city of Johannesburg, reflecting this to be an essential aspect of the commuting experience.

Historical research on transit users' perceptions has proven that traveling time has important implications on mode choice (Lagune-Reutler et al., 2016). Generally, commuters' perception is based upon the mode of transportation being able to follow its timetable; thus, any sudden deviations would lead to adverse emotions. Most commuters in South Africa use the railway system as an alternative to driving or using mini-bus taxis on the highways. Consequently, the introduction of the Gautrain was ideally meant to ease traffic congestion and as an alternative to driving or using commuter taxis on the highways. As can be seen, commuters' linguistic measure of time, has greatly improved, as there has been a steady growth in the number of positive comments on social media.

There is a need for public transportation systems to innovatively improve the societal benefits of their systems. This is because an imbalance in commuters' emotions over time could lead to them not using these modes as they would lose faith in the public transportation systems. Thus, in order for public transportation systems not to lose commuter loyalty they have to ensure they responded quickly, while reassuring the public that their grievance is being resolved, with minimum disruptions to service delivery.

9.4 Discussions

Through prediction, applications are now more widely used, although many obstacles still need to be tackled. Currently, knowledge gaps still exist about travel demand management (Giaimo et al., 2010). Hence, a bridge is still needed to link what is available (Web 2.0 data) and what could be done (planning). As no one model is sufficient for all situations, the need is essential to continuously develop and renew planning models. Using the experimental approach allowed the researcher to test the potential uses of Web 2.0 data. As illustrated in the results, having a comfortable commuting experience is essential for the respondents. Comfort can be dependent on soft, clean seats, a pleasant temperature, preferably having air-conditioning, and not many people in the vehicle. Its significance varies during the day, with trips during the off-peak showing a high demand for comfort in the various cities. Generally, innovative public transport commuters prefer the new buses with air-conditioning and a lower floor. Posts made during the peak time revealed the commuting experience as unpleasant, as these trips were described to be uncomfortable, too crowded, smelly, and airless. Several posts also articulated the need for availability of information. Furthermore, the ease of access to travel times, costs, and route information has a bearing on preferred mode choice. Also, innovative public transport is acknowledged as cheaper than private vehicles, but it does not appear as a key factor influencing people to change to public transport when considered in isolation from other factors which restrict that choice.

As a result, the chapter describes a set of tools that have been used to generate linguistic analysis from social media posts in order to visualize the societal impact of innovative public transport systems. The chapter has proved the value of using a large quantity of data because the standard error becomes greater if small amounts of records are used. In utilizing the model to analyze social media data, the issue of sampling becomes important. Thus, a large quantity of Web 2.0 data was used, as this improves the accuracy of the prediction, whereas a reduction would have an adverse effect. To determine the optimal quantity of data, however, one has to create a balance in what is available and the amounts of standard error acceptable for the results to still prove true.

Consequently, the analysis must be deeply embedded in the utilization of big data for the model to be effectiv An understanding of the correlation between the observed social media data points is of great importance. However, this also is dependent largely on the duration of data collection. Over time, more insight into the data can be gathered, using either a seasonal or yearly correlational analysis to identify the key factors that influence mobility.

9.5 Implications, Reflections, and Generalizability of Findings

As the purpose of the study varies, the applicability of the model should differ because scholars conduct research for numerous objectives. Using the assumption that people who post about the innovative public transportation systems are either existing or potential commuters reveals that the model does show the sphere of influence of these modes. However, in the real-world situation, this may not always be the case, as there is currently no clear way to ensure that everyone who posts about these modes is a current commuter. This can be resolved by correlation analysis between the data collected from Web 2.0 and commuter tag data from the respective innovative public transportation providers in the metropolitan cities.

9.6 Conclusion

The chapter has presented the potential benefits of utilizing Web 2.0 data in planning, while also analyzing how commuters engage in the various Web 2.0 platforms. Likewise, the National Land Transport Act of 2009 has also advocated for more sustainable means of promoting a reliable public transportation system. Hence the need exists to continuously use modern technologies to ensure that the current means of public transportation are able to cater to the ever-changing challenges of providing a reliable public transportation system. The chapter used a linguistic measure of analysis to determine factors that influenced commuters' behavior and perceptions about the innovative public transportation systems. This revealed commuters' emotions, that is whether their posts reflected negative or positive feels toward the innovative public transportation systems.

References

Arnold, K., Le Roux, A., & Hattingh, M. (2017). Impact of Gautrain stations on property prices and sales activity in the City of Johannesburg between 2006 and 2015. *South African Journal of Geomatics, 6*, 184–195.

Ceder, A. (2020). Urban mobility and public transport: Future perspectives and review. *International Journal of Urban Sciences*, 1–25.

Chai, C., Fan, J., Li, G., Wang, J., & Zheng, Y. (2019). Crowdsourcing database systems: Overview and challenges. In *2019 IEEE 35th International Conference on Data Engineering (ICDE)* (pp. 2052–2055). IEEE.

Di Pasquale, G., dos Santos, A. S., Leal, A. G., & Tozzi, M. (2016). Innovative public transport in Europe, Asia and Latin America: A survey of recent implementations. *Transportation Research Procedia, 14*, 3284–3293.

Gabdrakhmanova, N., & Pilgun, M. (2021). Cognitive systems, artificial neural networks and differential equations: Social media data. *Procedia Computer Science, 186*, 677–684.

Giaimo, G., Anderson, R., Wargelin, L., & Stopher, P. (2010). Will it work? Pilot results from first large-scale global positioning system-based household travel survey in the United States. *Transportation Research Record, 2176*(1), 26–34.

Huang, Q., & Wong, D. W. (2016). Activity patterns, socioeconomic status and urban spatial structure: What can social media data tell us? *International Journal of Geographical Information Science, 30*, 1873–1898.

Jiang, Y., Li, Z., & Ye, X. (2019). Understanding demographic and socioeconomic biases of geotagged Twitter users at the county level. *Cartography and Geographic Information Science, 46*, 228–242.

Lagune-Reutler, M., Guthrie, A., Fan, Y., & Levinson, D. (2016). Transit stop environments and waiting time perception: Impacts of trees, traffic exposure, and polluted air. *Transportation Research Record, 2543*(1), 82–90.

Levy Abitbol, J., Fleury, E., & Karsai, M. (2019). Optimal proxy selection for socioeconomic status inference on twitter. *Complexity, 2019.*

Marzano, G., Lizut, J., & Siguencia, L. O. (2019). Crowdsourcing solutions for supporting urban mobility. *Procedia Computer Science, 149***,** 542–547.

Mazumdar, S., & Thakker, D. (2020). Citizen science on Twitter: Using data analytics to understand conversations and networks. *Future Internet, 12*, 210.

Mbatha, S., & Gumbo, T. (2019). Establishing the state of spatial integratedness of innovative public transport systems in Gauteng, South Africa. In *REAL CORP 2019—IS THIS THE REAL WORLD? Perfect Smart Cities vs. Real Emotional Cities. Proceedings of 24th International Conference on Urban Planning, Regional Development and Information Society* (pp. 319–327). CORP—Compentence Center of Urban and Regional Planning.

McKay, T., Simpson, Z., & Patel, N. (2017). Spatial politics and infrastructure development: Analysis of historical transportation data in Gauteng-South Africa (1975–2003). Miscellanea Geographica. Regional Studies on Development, 21, 35–43.

Moyo, T., & Musakwa, W. (2016). Using crowdsourced data (Twitter & Facebook) to delineate the origin and destination of commuters of the gautrain public transit system in South Africa. *ISPRS Annals of Photogrammetry, Remote Sensing & Spatial Information Sciences, 3.*

Musakwa, W., & Gumbo, T. (2017). Impact of urban policy on public transportation in Gauteng, South Africa: Smart or dumb city systems is the question. Springer.

Porru, S., Misso, F. E., Pani, F. E., & Repetto, C. (2020). Smart mobility and public transport: Opportunities and challenges in rural and urban areas. *Journal of Traffic and Transportation Engineering (english Edition), 7*, 88–97.

Ryseck, B. (2019). South and Southern Africa—Are disruptive transport technologies ready? An exploration of journey planning tools for urban public transport systems in South Africa. In *Southern African Transport Conference.*

Van Gerven, M., & Bohte, S. (2017). Artificial neural networks as models of neural information processing. *Frontiers in Computational Neuroscience, 11*, 114.

Vanderschuren, M. J. W. A. (2006). Intelligent transport systems for South Africa: Impact assessment through microscopic simulation in the South African context.

Vermicelli, S., Cricelli, L., & Grimaldi, M. (2021). How can crowdsourcing help tackle the COVID-19 pandemic? An explorative overview of innovative collaborative practices. *R&D Management, 51*, 183–194.

Part VI
Conclusion

Chapter 10
Innovative Urban Public Transport Systems: An Evolving Continuous Expedition in Southern African Cities

Abstract The entirety of the transport issues discussed in this book highlight that innovations in urban public transport systems are a powerful force that continues to shape global cities in general and South African cities in particular. The nine preceding chapters have demonstrated the quest for adopting, implementing, and improving the integration of transport systems, spatially, functionally, and technologically in urban centers the world over. This is a great development, given the various transport challenges being experienced in cities of the global south. The innovations in urban public transport systems bring to the fore lessons on prudent strategies to improve the lived experiences of commuters and residents in global cities. However, the question becomes, what is the future of urban public transport systems in developing countries such as South Africa, and what is the role of the various stakeholders in creating an efficient transport system? The chapter wraps up the work that was presented in this book with a concise introduction of the main developments that have emerged from the preceding discussions. It then presents the implications and lessons learned from the new emerging developments and makes a case for the need for action by the various stakeholders who derive benefits from integrated urban public transport systems. The chapter also recommends the various strategies that could be adopted in ensuring that sustainable and integrated urban public transport systems are provided in other cities of the developing world. It uses as example, the efforts that have been deployed by the metropolitan cities in Gauteng province, thus replicating lessons learned from the current case studies. The chapter ends by suggesting new areas for further investigations that could be packaged into another volume and lastly, it provides a conclusion to the journey traveled.

Keywords Emerging developments · Multi-modal · Transit-oriented development · South Africa · Global south · Integration

10.1 Introduction

Urban public transport as a catalyst for urban development in the era of smart mobility is well recognized (Peprah et al., 2019). Typically, smart mobility describes city transport networks which are utilizing active travel modes; information and technology;

T. Gumbo et al., *Urban Public Transport Systems Innovation in the Fourth Industrial Revolution Era*, https://doi.org/10.1007/978-3-030-98717-6_10

energy-efficient, renewable forms of energy; or shared vehicles wherever possible, resulting in low carbon output per passenger journey (Dia, 2016; Namiot & Pokusaev, 2019). Integrated multi-modal-networked public systems have emerged as a smart mobility paradigm (Risimati & Gumbo, 2018). They use transfer potential to provide a maximal service for a reasonable and efficient operating budget and a genuinely feasible alternative to automobile travel within urban areas (Beaudoin & Lawell, 2018; Zambrano-Martinez et al., 2019). As mobility is the essence of modern life in urban areas, it creates serious social, economic, and environmental challenges. In a situation of growing car ownership, public transport services are facing tough competition from private automobiles. The continuous increase in the number of vehicles on the road network poses further threats to traffic movements. This is evidenced by traffic congestion, slower flow, more accidents, and waste of time, money, and efforts (Agyapong & Ojo, 2018; Wang et al., 2013).

For decades, a shift from mobility-centered to accessibility-centered transport and land-use planning has been advocated (Lyu et al., 2019; Straatemeier & Bertolini, 2020). This shift starts with a requirement for transportation primarily derived from people's need to reach their destination rather than for movement. It suggests that enhancing accessibility to desired destinations counts for the users of that transport system (Jäppinen et al., 2013; Saghapour et al., 2018). Transit-Oriented Development (TOD) can enhance accessibility by strengthening the integration between transport and land-use systems through relatively high density, mixed-use, cycling, and pedestrian-friendly development around transit stations and networks (Risimati & Gumbo, 2019). TOD characteristics are positively related to accessibility at the catchment level. Likewise, at the catchment level, the transport feature of the TOD system is highly related to accessibility, while for transit-oriented land-use patterns, the associations are much smaller (Lyu et al., 2020; Olaru & Curtis, 2015; Tong et al., 2018). If the transport system is kept unchanged, land-use policy relating to improving urban density, diversity, and pedestrian-friendly development is recognized as an effective tool to enhance the accessibility of the area.

Consequently, as discussed in the preceding chapters, past spatial planning practices have left Gauteng metropolitan cities with sprawling low-density areas of settlement, lacking viable public transport systems (Mbatha, 2019). Furthermore, the majority of working-class and poor citizens are still living on the fringes of the cities, commuting daily, often at considerable cost, over long distances to access work and economic opportunities (Ndwandwe, 2017). Private car use is also a significant driver of energy consumption and greenhouse gas emissions in cities (Moyo et al., 2021). It is thus critical for the chapter to assess developments that have emerged from the preceding discussions, present the implications and lessons learned from the study, and make a case for the need for action so as to derive benefits from integrated urban public transport systems. The chapter also recommends various strategies to ensure that sustainable and integrated urban public transport systems are provided in other cities of the developing world.

10.2 Emerging Developments

The nine preceding chapters have embraced new spatial plans, particularly the Corridor of Freedom based on transport-oriented development. The shape of the future Gauteng metropolitan cities will consist of well-planned transport arteries, with the Corridor of Freedom linked to interchanges, where the focus will be on mixed-use development. Corridors of Freedoms and development corridors are significant both as mobility spines from a high-level perspective within the Gauteng City Region, and public transport and pedestrian spines, supported by existing active street edges and land-uses. They have the potential to transform entrenched settlement patterns that have kept many marginalized communities at the outskirts of the cities, away from economic opportunities and access to jobs and growth. They guide future city growth toward areas best serviced by transit infrastructure and the full range of vibrant urban amenities and services. Therefore, the public will not have to use private transport but can opt for alternative means, including cycling, transit lanes, and pedestrian walkways. The Corridors of Freedom will transform entrenched settlement patterns which have shunted most residents to the cities' outskirts, away from economic opportunities and access to jobs and growth. Gone will be the days of being forced to rise at dawn to catch a train, bus, or taxi to a place of work. Families will have quality time with spouses and children, sharing meals in the evening.

To enable this vision, PRASA has completed a New National Plan which will guide infrastructure investment into specifically identified corridors. The City of Johannesburg Strategic Integrated Transport Plan Framework identifies a high-level public transport network for 2040, based on population growth, areas of employment growth, and projected densities. The Framework has identified several critical public transport corridors consolidating growth and development opportunities around existing and future public transport nodes, starting from the Corridors of Freedom linking Soweto, through the inner city, to Sandton (along Empire-Perth and Louis Botha Avenues) and linking Turffontein to the inner city. This will also include a focus on transit-oriented development nodes, including Gautrain, Rea Vaya (BRT), and Metrorail stations. Some of the public transport corridors will function as transit access areas, transporting large numbers of people from one part of the city to another. Other public transport corridors have the potential to grow into development access strips, with the opportunity to not only link mixed-use development nodes but to articulate public transit with housing, new employment activities, and social amenities, while optimizing investment capacities.

The proposed Corridors of Freedoms and development corridors are significant as mobility spines from a high-level perspective within the Gauteng City Region. They are also public transport and pedestrian spines, supported by existing active street edges and land-uses. They have the potential to transform entrenched settlement patterns that have kept many communities at the outskirts of the city, away from access to jobs and growth. They can also guide future city growth toward areas best serviced by transit infrastructure and the full range of vibrant urban amenities and services. The current initiatives intend to optimize development in and around high-intensity movement corridors to create more accessible opportunities for the residents of Johannesburg and economies of scale that are attractive to investors. Future growth around these access areas is envisaged as medium to high-rise residential developments growing around the transit nodes, gradually decreasing in height and density as development moves further away from the core. Social infrastructure, schools, clinics, police stations, and government offices will be strategically located to support the growing population. The future vision is mainly premised on theories and best practices around Transit Oriented Development (TOD). This seeks to create urban spaces with a vibrant mix of high-density residential developments, office, retail and recreational spaces within walkable precincts anchored by high-quality social amenities.

10.3 The Need for Action

The South African and Zimbabwe government's vision is to realize large-scale modal shifts from private to public transport to provide an integrated commuter transport system and access for all. This implies that a transportation system needs to provide what is attractive to users. Current public transport in the two countries, however, is generally regarded as being of a low standard. In the nine preceding chapters on the state of transport, an opinion poll indicated that the public is extremely dissatisfied with the quality and levels of public transport. Because of this, societal needs such as mobility and accessibility remain largely unaddressed. This is further evidenced by the high levels of private car usage as the dominant form of commuting and low levels of public transport use. Nationally, 38.48% of people access work as the driver or commuter within a private motor vehicle, 26.5% are made by minibus taxi (privately-owned public transport), and 21.1% walk all the way.

Transport policy in South Africa gives a valuable synopsis of what has transpired in the policy field. Since the 1996 White Paper on National Transport, several empowering legislative instruments have been adopted to promote integrated, efficient, reliable, convenient, and safe transport systems. These started with the 1996 National Constitution and the 1996 Green Paper on National Transport Policy. More recently, the Green Paper on the National Rail Policy was implemented in 2015. These helped identify methods of facilitating the planning, development, and management of enhanced transport systems in the country. The National Land Transport Act (Act 22 of 2000) was also a significant milestone for the national government, post the White Paper. This was followed by introducing the Integrated Rapid Public Transport Network (IRPTN) initiative, a more sophisticated approach to urban public transport systems (Van Ryneveld, 2008). As part of these innovations, the Gautrain and Bus Rapid Transit (BRT) was introduced, with the BRT being the commonly adopted form of public transport across the country. However, the previous recapitalization policies and programs did not do justice to the transformation of the minibus taxi industry. Instead, public funds were spent without even redressing the alarming safety concerns. This was also done without regard for modal integration. This continues to be the case, even with implementing innovative urban public transport systems consistent with the IRPTN national government strategy.

It is evident therefore from the study findings that without an integrated (multi-modal) urban public transport network, spatial integration will be hard to attain, regardless of how many technologically advanced modes of public transport are introduced. The appropriateness of a mode of public transport for the people and the purpose it serves is essential when devising strategies for modal integration. In this regard, the Gautrain was costly to implement, while also having high travel costs. It excludes the urban poor and previously disadvantaged communities, thus limiting its significance to connecting central economic nodes for business-related trips. Consequently, the need to integrate all (innovative and traditional) modes of public transport is encouraged as the backbone for the effective functioning of economic communities. When integrating various modes of public transport, it is essential to define the purpose of each type in the overall public transport network to ensure improvement of urban mobility. For example, the Gautrain may be designated to connect major economic nodes, while other modes transport commuters between residential and economic or other urban opportunities. A multi-modal public transport network will enhance spatial connectivity, densification, reduce travel ties and costs and thus enhance the efficiency of public transport systems. Through modal integration, infrastructure investments will be utilized more effectively to enable comprehensive public transport throughout municipal areas instead of unnecessarily duplicating services.

10.4 Working Toward Sustainable and Integrated Urban Public Transport Systems

Spatial inequality remains a defining characteristic of the settlement pattern of the cities in South Africa and Zimbabwe. The location and concentration of jobs do not match where people live which significantly contributes to inequality, as for many residents' access to economic opportunities is stifled by costly and distant commuting. Some of the highest densities of housing, the 'townships' inherited from apartheid spatial policies, are also some of the most deprived areas in South Africa, located far from areas of economic opportunities. In the City of Johannesburg, there are two major spatial structural discontinuities that are barriers to opportunity: the mining belt, which has become a symbol of north–south segregation, and the tracts of undeveloped land in the northeast of the city (such as Glen Austin/Austin view, Modderfontein, and Frankenwald suburbs). These form a spatial divide between the City of Johannesburg and its, neighbor, Ekurhuleni.

Post-apartheid housing delivery has further exacerbated apartheid spatial development patterns by building housing in areas far from economic activities. Land availability is the primary logic behind their location. Through car-oriented developments (for example, malls and gated residential estates and office parks), the private sector has further aggravated spatial segregation. Newer townhouses and cluster developments have relatively higher average residential density. These new developments are focused on private use and are frequently in single-use clusters with limited access to public transit infrastructure. They generally do not foster walkable neighborhoods and often have not been met with requisite public infrastructure. Some of the highest residential densities are some distance from the core economic activities. For example, Soweto, Orange Farm, Diepsloot, and Ivory Park reflect relatively high residential densities but are all limited in their land-use diversity.

Zimbabwe's high level of spatial inequality is reflected in the urban spatial form and high levels of securitization with the proliferation of developments with the CBD area only, thus leaving other areas marginalized. These are characterized by controlled street patterns that have moved from the historically open grid to the clustered cul-de-sac, loop, and 'lollipop' configuration, contributing to fragmentation and low levels of walkability. Continuing to meet development demand in this manner exacerbates existing socio-economic disparities in the cities. It also places significant pressure on the natural environment, reduces efficiency, and increases the cost of infrastructure provision (both to build and maintain over the long-term). Spatial concentration of formal jobs is also much higher than the spatial concentration of housing. This sharp concentration of jobs is an asset for the cities. It will feed economic growth if articulated with an efficient transportation network and increased housing opportunities in close proximity.

In interpreting the current Gauteng City Region morphology, Gauteng displays a unique inverted polycentricity, mainly inherited from its complex history. This structure is characterized, inter-alia, by peripheral or satellite nodes that are disproportionately large compared to main urban centers, and disconnected from them (inner city). Illogical density gradient residential areas also characterize Gauteng. For example, many high-density residential areas are located on the cities' outskirts, far from job and economic opportunities. This spatial contradiction translates into a job housing mismatch which exerts a significant impact on social exclusion and energy and carbon intensity (increasing travel time and travel distances from jobs to housing). It also impacts economic productivity (by jeopardizing agglomeration economies), with most commuter's flows being directed to the city center.

The South African government has prioritized improvement of transport systems through mega investment and strategic policy instruments. Moyo et al. (2021) further observe that the South African urban public transport system has reached a crucial stage, with major cities (supported by national and provincial governments) already geared up to implement innovative public transport infrastructure. In this regard, metropolitan cities in South Africa seem to be at the center of innovative transport systems initiatives, while other cities and towns have lagged. This is understandable, given the population concentration and significant economic activities in metropolitan cities. At present, the study findings of the preceding chapters also reveal that most commuting and cycling operations in Gauteng Province are spatially disintegrated, with inadequate to no sharing of their infrastructure. There is no adequate cycling infrastructure in most public transit stations, and these are not easily accessible for non-motorized transport. This lack of integration of commuting and cycling infrastructure impedes the realization of a multi-modal system, as residents are unable to utilize cycling for the first and last mile of commuting trips. Identifying an urban public transportation catchment area also is essential for Metropolitan developing cities within Gauteng. Such areas can be used as a tool for planning infrastructural upgrades and predicting potential public transport ridership while assessing the impacts of investments in transport planning.

Integrated public transport planning improves a cities' connectivity, providing a better mobility service, shorter journeys for commuters, and also brings people and places closer together. The Gauteng City Region should therefore consider integrating infrastructures and operations of urban public transport systems and creating easy connections with non-motorized transport (such as walking and cycling). Well-integrated public transport modes will increase their usage by commuters. As commuters get more value for money, they will consider public transport as a more convenient mobility option. One of the main benefits is that integrated transport systems foster social equality, providing access to services, jobs, education, and entertainment, and access to the whole province. Peoples' use of more sustainable modes of transport can reduce congestion, emissions, travel times, and if appropriately managed, even road accidents.

Because the different spheres of government have implemented their projects independently, spatial connectivity through multi-modal urban public transport networks is still not effectively pursued. This has led to the creation of modes of public transport that are operationally disintegrated. Even different types of innovative urban public transport systems operate independently of other existing forms or modes of urban transport. Thus, duplication of multiple modes of public transport in the same geographical area without making any difference is inevitable. It should be acknowledged that minibus taxis transport most commuters from previously disadvantaged communities and therefore should form an integral part of the public transport modal integration. The minibus taxi industry is still wholly disregarded in the current transport innovation, despite being responsible for transporting sizable numbers of commuters from disadvantaged communities. The previous recapitalization program did not do justice to the transformation of the minibus taxi industry. Instead, public funds were spent without even redressing the alarming safety concerns. This was done without regard for modal integration.

It is evident from the study findings that without a multi-modal urban public transport network, spatial integration will be hard to attain, regardless of how many technologically advanced modes of public transport are introduced. The appropriateness of a mode of public transport for the people and its purpose is essential when devising strategies for modal integration. In this regard, the Gautrain was costly to implement while also having high travel costs. It excludes the urban poor and previously disadvantaged communities, thus limiting its significance to connecting major economic nodes for business-related trips. Thus, the need to integrate all (innovative and traditional) modes of public transport is encouraged as the foundation for the effective functioning of economic communities. When integrating varied modes of public transport, it is essential to define the purpose of each type of public transport in the overall public transport network to achieve sustainable mobility. For example, the Gautrain may be designated to connect major economic nodes, while other modes transport commuters between residential and economic or other urban opportunities. A multi-modal public transport network will enhance spatial connectivity, densification, reduce travel ties and costs and thus enhance the efficiency of public transport systems. Through modal integration, infrastructure investments will be utilized more effectively to enable comprehensive public transport throughout municipal areas instead of unnecessarily duplicating services.

10.5 Implications, Reflections, and Generalizability of Findings

Results support public transport as a sustainable means for reducing congestion, saving the use of public space, alleviating environmental issues caused by automobiles, and providing affordable transportation, especially for low-income groups. Determining urban public transportation service coverage or catchment area is essential in predicting potential public transport ridership and assessing the impacts of investments in transit planning. Because most public transport commuters are pedestrians or cyclists at the beginning or end of their trips, a public transport catchment area is conventionally determined by the willingness distance or distance threshold to walk. Thus, based on the discussion above, three research questions are raised:

- How can innovative public transportation systems be integrated with traditional modes?
- What infrastructure will then be required at interchange zones?
- Lastly, how can innovative public transportation solve mobility challenges while being sustainable?

To answer these questions, this chapter suggests that future research should examine analytical features, including travel distance and duration of walking and cycling trips to access urban public transport services and the potential of green mobility systems. The urban public transport service coverage should be ascertained by the number of households and employment within transit catchment areas. The connectivity and quality of urban mobility networks should be adopted by calculating non-motorized accessibility to transit. Through these methods, the raised questions can be addressed.

10.6 Conclusion

This chapter has wrapped up the work presented in the nine preceding chapters of this book. It has assessed developments emerging from the preceding discussions, presented the implications and lessons learned from the study, and made a case for the need for action to derive benefits from integrated urban public transport systems. The study has embraced new spatial plans to identify mobility spines from a high-level perspective within the Gauteng City Region. The preceding discussions reveal that commuting operations in Zimbabwe and South African cities are spatially disintegrated, with inadequate to no sharing of infrastructure. Without a multi-modal urban public transport system, spatial integration will be hard to attain. Thus, the integration of all (innovative and traditional) modes of public transport is encouraged as the backbone of the effective functioning of economic communities. It is suggested that future research should examine analytical features. The urban public transport service coverage should be investigated by the number of households and employment within transit catchment areas. Spatial boundaries of pedestrian-transit and bicycle-transit catchment areas should be determined based on identified distance thresholds and non-motorized accessibility to public transport.

References

Agyapong, F., & Ojo, T. K. (2018). Managing traffic congestion in the Accra Central Market, Ghana. *Journal of Urban Management, 7*(2), 85–96.

Beaudoin, J., & Lawell, C. Y. C. L. (2018). The effects of public transit supply on the demand for automobile travel. *Journal of Environmental Economics and Management, 88*, 447–467.

Dia, H. (2016). The real-time city: Unlocking the potential of smart mobility. In *Australasian Transport Research Forum 2016, Proceedings* (pp. 16–18), Melbourne, Australia.

Jäppinen, S., Toivonen, T., & Salonen, M. (2013). Modelling the potential effect of shared bicycles on public transport travel times in Greater Helsinki: An open data approach. *Applied Geography, 43*, 13–24.

Lyu, G., Bertolini, L., & Pfeffer, K. (2019). How transit-oriented land-use is related to accessibility? A study in Beijing. *Transportation Research Procedia, 41*, 52–54.

Lyu, G., Bertolini, L., & Pfeffer, K. (2020). How does transit-oriented development contribute to station area accessibility? A study in Beijing. *International Journal of Sustainable Transportation, 14*(7), 533–543.

Mbatha, S. G. (2019). *Investigating the possibility of electronic intermodality and interoperability of innovative urban public transport systems in the city of Tshwane*. University of Johannesburg (South Africa).

Moyo, T., Kibangou, A. Y., & Musakwa, W. (2021). Societal context-dependent multimodal transportation network augmentation in Johannesburg, South Africa. *Plos One, 16*(4), e0249014.

Namiot, D., & Pokusaev, O. (2019). On mobility patterns in smart city. In *CEUR Workshop Proceedings* (pp. 19–28).

Ndwandwe, B. (2017). *The application of principal component analysis in examining the impact of public transport systems on socio-economic transformation in the city of Tshwane*. University of Johannesburg (South Africa).

Olaru, D., & Curtis, C. (2015). Designing TOD precincts: Accessibility and travel patterns. *European Journal of Transport and Infrastructure Research, 15*(1), 6–26.

Peprah, C., Amponsah, O., & Oduro, C. (2019). A system view of smart mobility and its implications for Ghanaian cities. *Sustainable Cities and Society, 44*, 739–747.

Risimati, B., & Gumbo, T. (2018). Exploring the applicability of location-based services to delineate the state public transport routes integratedness within the city of Johannesburg. *Infrastructures, 3*(3), 28.

Risimati, B., & Gumbo, T. (2019). Mapping the spatial integration of motorised and non-motorised transport infrastructures: A case study of the city of Johannesburg. In *Is this the real world?*

Perfect smart cities vs. real emotional cities. Proceedings of REAL CORP 2019, 24th International Conference on Urban Development, Regional Planning and Information Society (pp. 503–510). CORP—Competence Center of Urban and Regional Planning.

Saghapour, T., Moridpour, S., & Thompson, R. G. (2018). Enhancing active transport demand modelling by incorporating accessibility measures. *Cities, 78*, 206–215.

Straatemeier, T., & Bertolini, L. (2020). How can planning for accessibility lead to more integrated transport and land-use strategies? Two examples from the Netherlands. *European Planning Studies, 28*(9), 1713–1734.

Tong, X., Wang, Y., Chan, E. H., & Zhou, Q. (2018). Correlation between transit-oriented development (TOD), land use catchment areas, and local environmental transformation. *Sustainability, 10*(12), 4622.

Van Ryneveld, P. (2008). *15 year review of public transport in South Africa with emphasis on metropolitan areas*. Unpublished paper commissioned for the Fifteen Year Review.

Wang, C., Quddus, M., & Ison, S. (2013). A spatio-temporal analysis of the impact of congestion on traffic safety on major roads in the UK. *Transportmetrica A: Transport Science, 9*(2), 124–148.

Zambrano-Martinez, J. L., Calafate, C. T., Soler, D., Lemus-Zúñiga, L. G., Cano, J. C., Manzoni, P., & Gayraud, T. (2019). A centralized route-management solution for autonomous vehicles in urban areas. *Electronics, 8*(7), 722.

The manufacturer's authorised representative in the EU is Springer Nature Customer Service Centre GmbH, Europaplatz 3, 69115 Heidelberg, Germany. If you have any concerns regarding our products, please contact ProductSafety@springernature.com

Printed and bound by CPI Group (UK) Ltd, Croydon, CR0 4YY
30/09/2026
02225493-0001